I0489269

NUREG-1889

RASCAL 3.0.5 Workbook

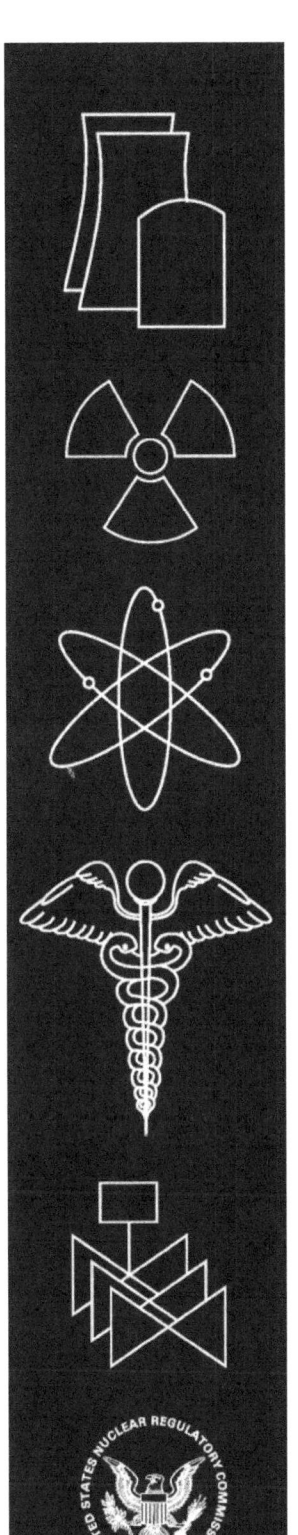

Manuscript Completed: September 2007
Date Published: September 2007

Prepared by
G.F. Athey[1]
S.A. McGuire[2]
J.V. Ramsdell, Jr.[3]

[1]Athey Consulting
P.O. Box 178
Charles Town WV 25414-0178

[2]Office of Nuclear Security and Incident Response
U.S. Nuclear Regulatory Commission
Washington, DC 20555-0001

[3]Pacific Northwest National Laboratory
P.O. Box 999
Richland, WA 99352

**Prepared for
Office of Nuclear Security and Incident Response
U.S. Nuclear Regulatory Commission
Washington, DC 20555-0001**

Abstract

The code currently used by NRC's emergency operations center for making dose projections for atmospheric releases during radiological emergencies is RASCAL version 3.0.5 (Radiological Assessment System for Consequence AnaLysis). This code was developed by NRC. RASCAL 3.0.5 evaluates releases from: nuclear power plants, spent fuel storage pools and casks, fuel cycle facilities, and radioactive material handling facilities.

The workbook contains problems designed to familiarize the user with the RASCALsoftware through hands-on problem solving. The workbook is primarily for use by students in a RASCAL training course supervised by a qualified instructor.

Table of Contents

List of Figures

List of Tables

Acknowledgments

The authors would like to acknowledge all the students who have used the workbook over the years. Your comments and suggestions have been invaluable.

Introduction

This workbook is designed to teach you how to use RASCAL through hands-on problem solving. The problems described within were developed using the December 2006 release of RASCAL v3.0.5. The workbook may be used with other revisions of the software. However, the results from working the problems may appear different from those in the workbook.

Attempt to do each problem using the computer and answer all the associated questions. Blank tables are provided for you to record your results. A description of how the problem may be solved and a discussion of the results are included with each problem.

You must understand the risks and uncertainties inherent in dose projection models before using this tool.

Contact Information

Please direct any questions, suggestions, corrections or comments concerning this workbook to the following:

Stephen A. McGuire
Office of Nuclear Security and Incident Response
U.S. Nuclear Regulatory Commission
Washington, DC 20555-0001
sam2@nrc.gov

George F. Athey
Athey Consulting
PO Box 178
Charles Town, WV 25414-0178
atheyconsulting@frontiernet.net

Lou Brandon
Office of Nuclear Security and Incident Response
U.S. Nuclear Regulatory Commission
Washington, DC 20555-0001
lkb1@nrc.gov

Registration

To become a registered RASCAL user and receive program updates automatically, e-mail your contact information to George Athey.

Distribution

RASCAL was developed by the NRC using entirely government funds. As such, it has no copyrights or commercial restrictions. RASCAL can be freely copied, installed, and distributed. It contains no licensed components and requires no registration, activation, or royalties.

For information on how to obtain the RASCAL computer code or a PDF version of this workbook, contact George Athey.

1 What Is RASCAL?

RASCAL, which stands for Radiological Assessment System for Consequence AnaLysis, is the software developed and used by the U. S. Nuclear Regulatory Commission (NRC) Emergency Operations Center to estimate projected doses from radiological emergencies.

RASCAL consists of two main tools: the Source Term to Dose model and the Field Measurement to Dose model.

Source Term to Dose

Use the Source Term to Dose model when you want to enter information about plant conditions or accident conditions in order to estimate projected radiation doses from a plume to people downwind. The Source Term to Dose model will first generate a time-dependent "source term". This is the release rate for each radionuclide from the facility as a function of time. This time-dependent release rate (the "source term") then provides the input to an atmospheric dispersion and transport model.

The atmospheric dispersion and transport model estimates radionuclide concentrations downwind, both in the air and on the ground due to deposition. The calculated concentrations are then use to estimate projected doses. The dose pathways are: cloudshine from the plume, inhalation from the plume, and groundshine from deposited radionuclides (assuming 4 days of exposure to groundshine).

Field Measurement to Dose

Use the Field Measurement to Dose model when the plume-phase of the accident is over and you are in the intermediate-phase. The Field Measurement to Dose model estimates doses based on measurements of the radionuclide activity in the environment. The model will use the ground concentration of radionuclides to calculate intermediate-phase doses.

Two additional RASCAL Tools

Use the Decay Calculator to evaluate the radioactive decay of a radionuclide mixture over a user specified decay time.

Use the Nuclide Data Viewer to examine the contents of the RASCAL radionuclide database. This database contains information on individual radionuclides such as half-life, dose conversion factors, and inventories.

2 Getting Started with RASCAL

RASCAL is started by selecting in Windows:

Start | All Programs | RASCAL v3.0.5

This will display the opening screen with version number and date of release.

Click the **OK** button to continue to the main RASCAL access screen (shown below).

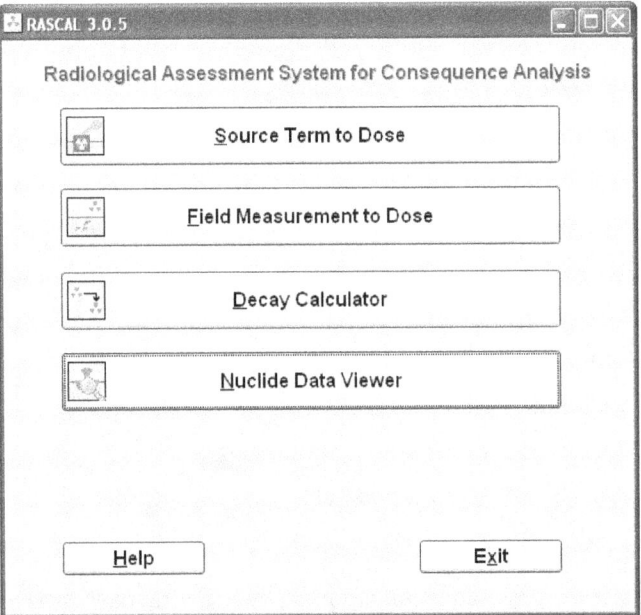

The four big buttons start the respective tools. Since the first section of the workbook deals primarily with the **Source Term to Dose** model, click that button.

Basic Features of the User Interface

This is the main screen of the **Source Term to Dose** model as seen when first starting up the program.

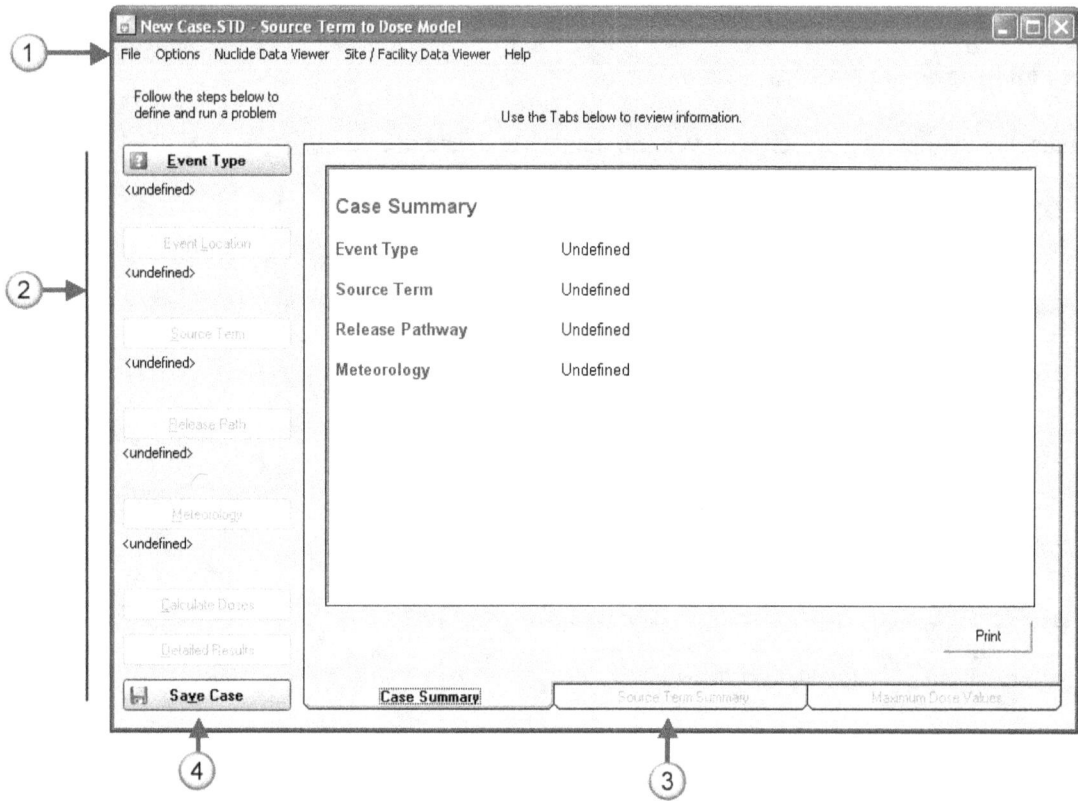

1. The menu bar at the top provides the standard Windows functions such as file management and help as well as access to two data viewers.

    The **Nuclide Data Viewer** can be used to examine the contents of the RASCAL radionuclide database. The database contains nuclide specific information such as inventories, dose conversion factors, and half-lives.

    The **Site / Facility Data Viewer** can be used to display data on each of the facilities in the RASCAL database. This includes geographic coordinates and a brief location description plus details about the facility. For reactors it includes parameters used by the model such as volumes, pressures, and leak rates.

2. The main input to the program is controlled via the buttons on the left side; starting from the top and working downward. The buttons are blue with a question mark when they are available but the topic area has not yet been defined. The buttons display a checkmark when the step has been completed.

3. The right portion of the screen contains a set of three tabs. These tabbed areas provide easy viewing of the inputs and some of the generated results. Initially, only the case summary tab is active and updates as the problem is defined. After a set of calculations has been completed, the **Source Term Summary** and **Maximum Dose Values** tabs become active to show results. More results are accessed via the **Detailed Results** button.

4. The **Save Case** button is used to create a single file containing all the inputs and results from a single modeling run. (The save file is really just a zipped archive containing all the input and output files generated during the building and running of the case. You could use an unzip utility to extract and examine files in the saved case.)

All the user inputs are remembered after the calculations. Thus, it is very easy to go back and change a single input and recalculate without having to completely redo the case inputs.

3 How RASCAL Calculates a Reactor Source Term

The RASCAL source term calculations for nuclear power plant accidents are based largely on the methods described in NUREG-1228 (McKenna and Giitter 1988).

The basic method first determines activity present by radionuclide. Then, the method multiplies the activity present by the fraction of the material present that is available for release. This product may be multiplied by a reduction factor, if appropriate. Examples of reduction mechanisms are filters, plateout, and removal by containment sprays. This product is then multiplied by a leak rate.

For example, this calculation could describe the release of radionuclides from a reactor core to a containment building or it could describe the release of radionuclides from a containment building to the environment.

In general, releases will in most cases occur over an extended period of time. The release rate will generally vary as a function of time. The time dependence of the release is treated by estimating the activity release rate averaged over discrete time steps that have a duration of 15 minutes.

Core inventories

For nuclear power plant source terms based on core damage, the radionuclide inventories assumed to be in the reactor core are shown in Table 1.

The values in Table 1 are for low-enriched uranium fuel. The values were derived from Table 2.2 of NUREG-1228 (McKenna and Giitter 1988). The derivation was done by dividing the fission product inventories in NUREG-1228 by 3 to convert from Ci/MWe to Ci/MWt, and then rounding to 2 significant figures. The inventories of nuclides with a half-life of more than one year were scaled up from a burnup of 18,000 to 30,000 MWD/MTU. The inventories in NUREG-1228 were derived from the Reactor Safety Study, WASH-1400. Table VI 13-1 of WASH-1400 ranked the nuclides in the core by importance to early health effects. The nuclides with greater importance for early health effects were included in the NUREG-1228 core inventory list. In addition, NUREG-1228 added noble gases that had lesser importance for early health effects because noble gases are the most likely group of fission products to be released to the environment by a nuclear power plant accident. According to Figure 2-4 of NUREG-1228, the iodine and tellurium nuclides contribute almost two-thirds of the bone marrow dose for a major nuclear power plant release. Krypton, cesium, strontium and barium are the other major contributors.

Table 1 Nuclear Power Plant Core Inventory During Operation for Low Enriched Uranium Fuel (30,000 MWD/MTU burnup)

Nuclide	Core Inventory Ci/MW(t)	Nuclide	Core Inventory Ci/MW(t)
Ba-140	5.30e+04	Ru-103	3.70e+04
Ce-144	2.80e+04	Ru-106	1.33e+04
Cs-134	4.17e+03	Sb-127	2.00e+03
Cs-136	1.00e+03	Sb-129	1.10e+04
Cs-137	2.67e+03	Sr-89	3.10e+04
I-131	2.80e+04	Sr-90	2.00e+03
I-132	4.00e+04	Sr-91	3.70e+04
I-133	5.70e+04	Te-129m	1.80e+03
I-134	6.30e+04	Te-131m	4.00e+03
I-135	5.00e+04	Te-132	4.00e+04
Kr-85	3.17e+02	Xe-131m	3.30e+02
Kr-85m	8.00e+03	Xe-133	5.70e+04
Kr-87	1.60e+04	Xe-133m	2.00e+03
Kr-88	2.30e+04	Xe-135	1.10e+04
La-140	5.30e+04	Xe-138	5.70e+04
Mo-99	5.30e+04	Y-91	4.00e+04
Np-239	5.50e+05		

Reference: Derived from NUREG-1228, Table 2.2 (McKenna and Giitter 1988) which in turn derived its table from WASH-1400

The inventories in Table 1 are based on a burn-up of 30,000 MWD/MTU. RASCAL adjusts the inventory of radionuclides that have a half-life exceeding one year to account for burnup. The inventory for the specified actual burnup, INV_{ACTUAL}, is calculated only for nuclides with a half-life of more than one year using the equation below. There is no burnup adjustment for nuclides with a half-life less than one year.

$$INV_{ACTUAL} = INV_{30,000} \times \frac{BURNUP_{ACTUAL}}{30,000\,MWD\,/\,MTU}$$

If the reactor is shut down prior to the start of the release, the radionuclide inventories are adjusted to account for radiological decay and ingrowth. In addition, at the end of each time step, the activities of the nuclides present are adjusted to account for radiological decay and ingrowth. The minimum activity of a nuclide allowed in a source term time step is 10^{-15} Ci.

Coolant Inventories

RASCAL uses coolant inventories for some accident types. The concentrations that RASCAL uses for normal coolant are given in Table 2.

Table 2 Radionuclide concentrations in reactor coolant

Nuclide	PWR coolant concentration	BWR coolant concentration	Nuclide	PWR coolant concentration	BWR coolant concentration
	Ci/g	Ci/g		Ci/g	Ci/g
Ba-140	1.30e-08	4.00e-10	Mo-99	6.40e-09	2.00e-09
Ce-144	4.00e-09	3.00e-12	Np-239	2.20e-09	8.00e-09
Co-58	4.60e-09	1.00e-10	Ru-103	7.50e-09	2.00e-11
Co-60	5.30e-10	2.00e-10	Ru-106	9.00e-08	3.00e-12
Cs-134	3.70e-11	3.00e-11	Sr-89	1.40e-10	1.00e-10
Cs-136	8.70e-10	2.00e-11	Sr-90	1.20e-11	7.00e-12
Cs-137	5.30e-11	8.00e-11	Sr-91	9.60e-10	4.00e-09
H-3	1.00e-06	1.00e-08	Tc-99m	4.70e-09	2.00e-09
I-131	2.00e-09	2.20e-09	Te-129m	1.90e-10	4.00e-11
I-132	6.00e-08	2.20e-08	Te-131m	1.50e-09	1.00e-10
I-133	2.60e-08	1.50e-08	Te-132	1.70e-09	1.00e-11
I-134	1.00e-07	4.30e-08	Xe-131m	7.30e-07	0
I-135	5.50e-08	2.20e-08	Xe-133	2.90e-08	0
Kr-85	4.30e-07	0	Xe-133m	7.00e-08	0
Kr-85m	1.60e-07	0	Xe-135	6.70e-08	0
Kr-87	1.70e-08	0	Xe-138	6.10e-08	0
Kr-88	1.80e-08	0	Y-91	5.20e-12	4.00e-11
La-140	2.50e-08	4.00e-10			
Mn-54	1.60e-09	3.50e-11			

Reference: ANSI/ANS 18.1-1999.

Those normal coolant concentrations are taken from ANSI/ANS 18.1-1999. During steady-state conditions, iodine and other fission products may escape from fuel rods having clad defects and enter the reactor coolant system. Since the internal pressure in the fuel rod is balanced with the coolant pressure outside the fuel rod during steady-state conditions, the rate of escape is low. The fission products that do escape into the reactor coolant system are continually removed by the reactor coolant system purification cleanup resulting in a low equilibrium concentration.

However, if a reactor transient causes the pressure of the reactor coolant system to decrease rapidly, the escape rate from fuel rods can increase and cause a temporary increase, or "spike," in the coolant concentrations. There is also a belief that coolant water can enter fuel rods through cladding defects. If the reactor coolant system pressure suddenly decreases, this water could leach off iodine and cesium salts deposited on the inner clad surfaces, increasing the iodine and cesium available for escape during the transient.

Using RASCAL: Assessing a PWR Core Damage Accident

This section of the workbook provides a detailed look at using RASCAL to develop a protective action recommendation for a PWR core damage accident. It is a walk through the entire process from start to finish.

The problems in this section should be worked in sequence as each depends on the previous work.

4 Source Term and Release Pathway

Purpose

To learn how to select source terms, release pathways and reduction mechanisms.

Discussion

Here are a few facts that you should know:

- Almost all the radioactive material at a nuclear power plant is in the fuel rods. Therefore, only major damage to fuel rods can cause a large release.

- The only plausible way to cause major damage to fuel rods is to remove the water covering the rods so that the heat produced by radioactive decay can no longer be removed effectively.

- If water no longer covers the fuel rods, the fuel cladding will first start to melt. Later, the fuel itself will start to melt. If the fuel remains uncovered, the molten fuel will eventually melt through the reactor pressure vessel.

- Any accident that leaves the fuel rods uncovered is a very serious accident that will cause large amounts of fission products to be released from the reactor core.

- It is a lot easier to predict when the uncovering of a reactor core will occur and how long the uncovering will last than it is to predict how much core damage will occur (for example, what percent of the fuel will melt). Therefore, RASCAL uses the time that the reactor core is uncovered as its primary method to assess offsite consequences of accidents involving core damage.

- When you use the "time core is uncovered" source term type, RASCAL will calculate the time-dependent release of fission products to the containment. The first fission products released from the core are noble gases and volatile elements like iodine and cesium. Non-volatile materials do not come out until later.

- The "time core is uncovered" source term option of RASCAL is likely to overestimate the speed and magnitude of the release and thus also overestimate the projected radiological doses. RASCAL users should inform decision makers of that fact.

Problem

You are provided with the following information:

- Arkansas Nuclear One, Unit 1 had been operating at full power

- At 10:00 A.M. the reactor scrammed due to a major rupture in the primary coolant system (loss-of-coolant accident)

- Operators believe the core may become uncovered at 1:00 P.M. Further, they are concerned they may not be able to provide enough makeup water to recover the core.

- Plant operators report they are unable to activate the containment spray system. However, they expect the containment to remain intact and any release to the atmosphere will be at the design leak rate.

Use RASCAL to estimate potential consequences from the accident.

If needed, start the Source Term to Dose model from the main RASCAL screen (see page 3). Then, input the information provided into the Event Type, Event Location, Source Term, and Release Path screens of the model.

Inputs: Step-by-step

We start with the main screen showing an undefined case and waiting for the event type to be selected.

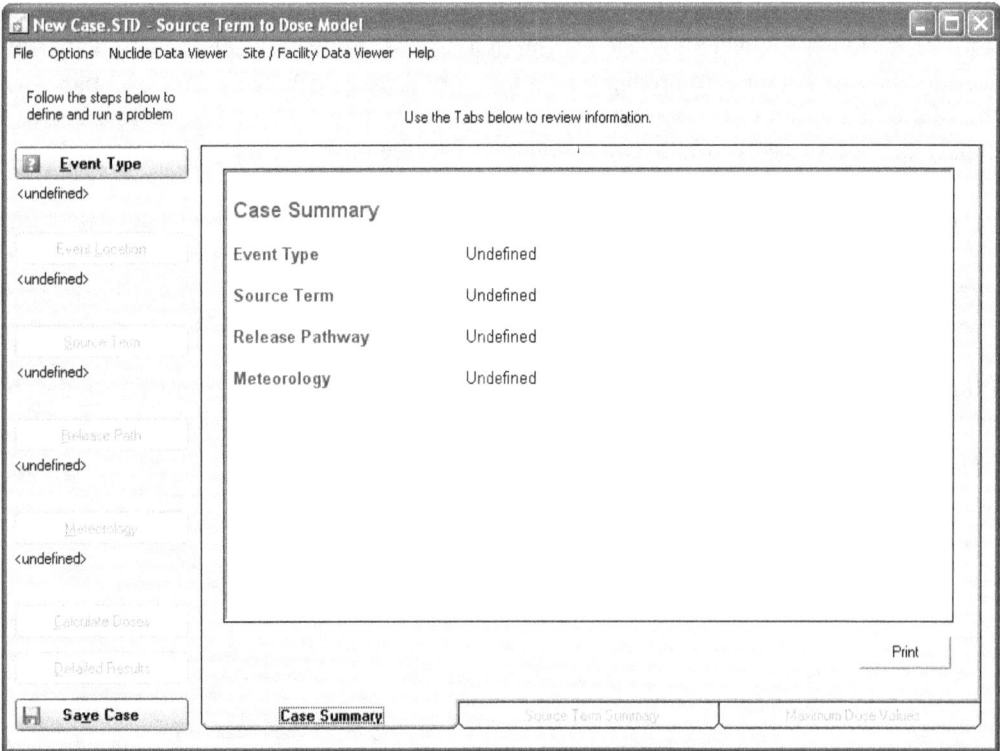

Event Type

Selection of the event type is easy. First, click the **Event Type** button. Since the problem is at Arkansas Nuclear and is a reactor accident - choose the option **Nuclear Power Plant**. Making this selection will limit the choices you have further down the line.

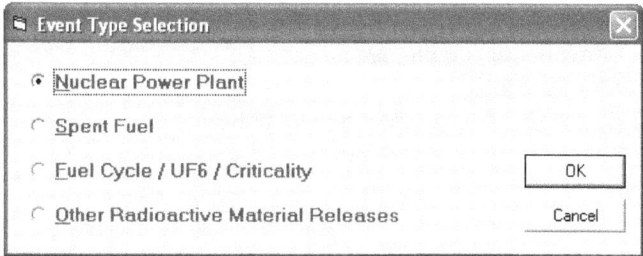

Click the **OK** button to return to the main screen. The **Event Type** button now has a checkmark indicating that the step has been completed. Also, the **Event Location** button color has now changed from grey to blue indicating it is active.

Event Location

Specification of the event location is also easy in this case. Click the **Event Location** button to bring up the **Location and Plant Parameters** screen.

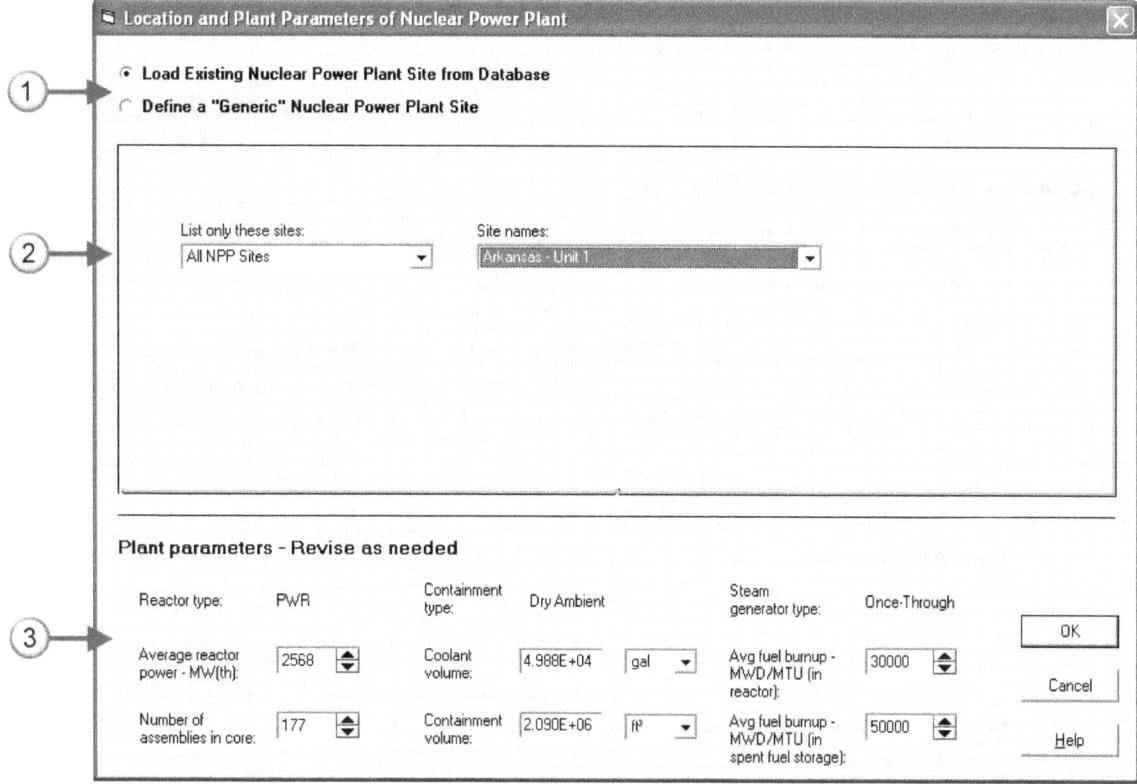

1. Select the option to **Load Existing Nuclear Power Plant Site from Database**

   RASCAL contains a database with information about all the commercial nuclear power plants in the U.S. The database contains unit specific information such as power and volumes as well as location and population information. The second option (**Define a "Generic" Nuclear Power Plant Site**) is available if needed. It can be used to define a site anywhere in the world.

2. Select **Arkansas - Unit 1** from the list of available site names

   The drop-down list has all the nuclear power plants available in the database.

3. Leave all the plant parameters unchanged

   This section displays some of the plant specific data from the database. The values can be changed if needed. However, the problem states that the reactor had been at full power so leave the average power set to the default. Leave the average fuel burnup at the default value of 30,000 MWD/MTU because better information was not provided.

   The average reactor power has units of megawatts thermal (MWt). The thermal power is a direct measure of the energy produced by fission. Therefore, the fission product inventory will be proportional to the thermal power. Changing the reactor thermal power level affects the release quantity of all radionuclides.

   (The electric power is the thermal power times the electrical generating efficiency of the plant. Typical efficiencies are usually between 30% and 35%, giving a rough conversion factor of 3 between MWe and MWt. All plants have slightly different efficiencies and the efficiency of a single plant can vary depending on operating conditions. Thus, fission product inventory is not exactly proportional to electric power output. It is usually close, but not exact.)

   Fuel burnup is a measure of how much fission energy has been produced by the fuel elements that are currently in the core. The fission energy released per unit mass of the fuel has units of megawatt-days per metric ton of uranium (MWD/MTU).

   A brand new core will have zero burnup. After several reloads of reactor fuel (generally replacing about one-third of the core per reload), the burnup will typically be about 20,000 MWD/MTU at the start of the cycle and will steadily increase to about 40,000 MWD/MTU at the end of cycle when the reactor shuts down to replace old fuel with fresh new fuel. The older fuel elements that are replaced will typically have a burnup of about 50,000 MWD/MTU. The reactor control room can determine the current fuel burnup at any point in the cycle, although this will not a have high priority during an actual emergency.

   Radionuclides with relatively short half-lives (less that a couple of weeks) reach a steady-state equilibrium activity (quantity) fairly rapidly. Their activity in the core is largely unaffected by burnup.

   But, the activities of long-lived radionuclides (half-life greater than one year) steadily increase as the reactor core produces power. Their activities never reach equilibrium. RASCAL uses average core burnup to adjust the core inventory for the long-lived radionuclides in the core fuel that never reach equilibrium. All long-lived radionuclides in the core inventory are scaled as needed based upon the actual burnup value entered.

   Click **OK** to return to the main screen.

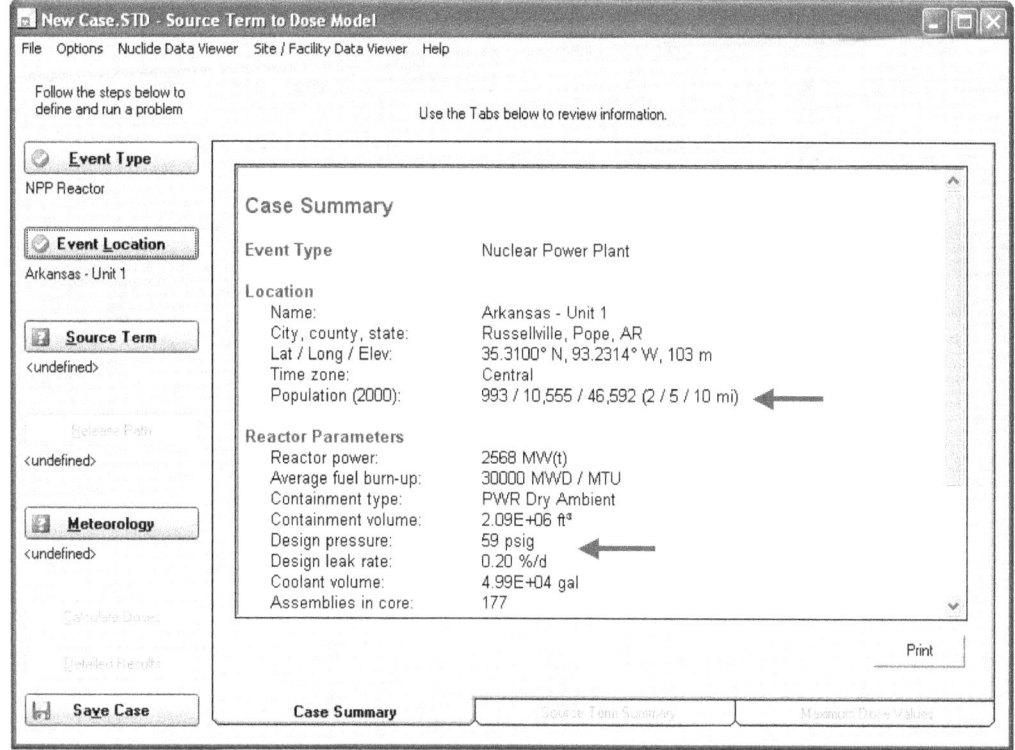

Note that the case summary is filled in as you proceed through the definition of the problem. This includes the information you enter as well as supplemental information pulled from the RASCAL databases. For example, the 2000 Census population at 2, 5, and10 mile radii from the plant are shown. Also shown are, the design pressure and design leak rate.

Source Term

Now select the **Source Term** button. This will display the **Source Term Options for Nuclear Power Plant** screen. All the available source term types for a nuclear power plant accident are listed.

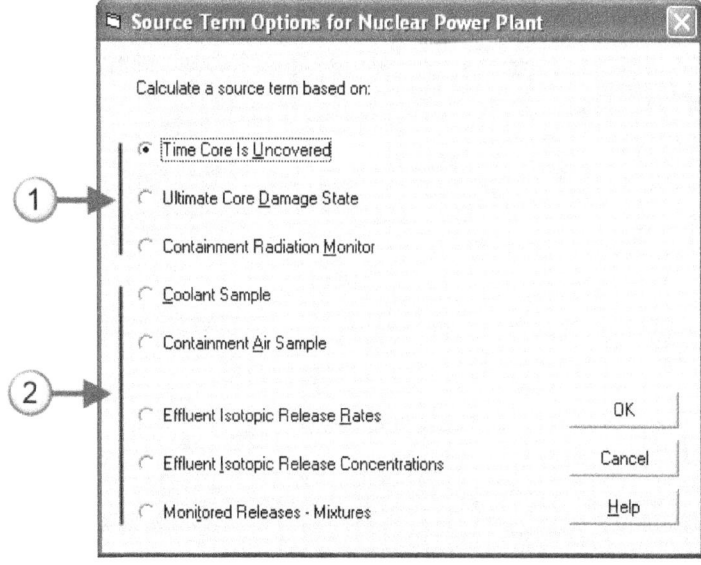

1. The top three options base the source term on an estimate of the amount of core damage.

2. The bottom five options base the source term on some form of measurement.

Select the **Time Core Is Uncovered** option on this screen and then click **OK** to proceed.

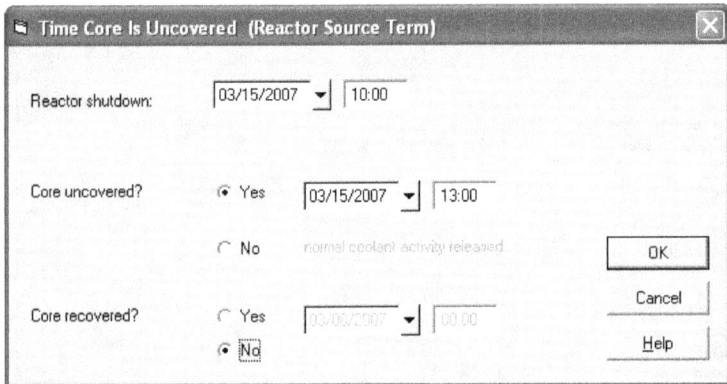

Enter the time of the reactor shutdown (**10:00**) and the time the core was uncovered (**13:00**).

There are two important things to understand here about times in RASCAL. First, RASCAL uses a 24 hour clock (instead of A.M. and P.M.). That is why we entered 13:00 instead of 01:00. Second, you should always work in *site time*. All information received from the plant will be in their local time. Trying to convert time zones can be confusing in an emergency and is not necessary if all parties have agreed to work with site time.

Leave the option for **Core recovered** set to **No**. Click **OK** to complete the source term definition.

Release pathway

Now select the **Release Pathway** button on the main screen. This will display the **PWR Release Pathways** screen.

Each source term type will have one or more available release pathway. As shown below, there are three release pathway options for a PWR. Each option has an associated containment schematic which uses red lines to show pathways to the atmosphere.

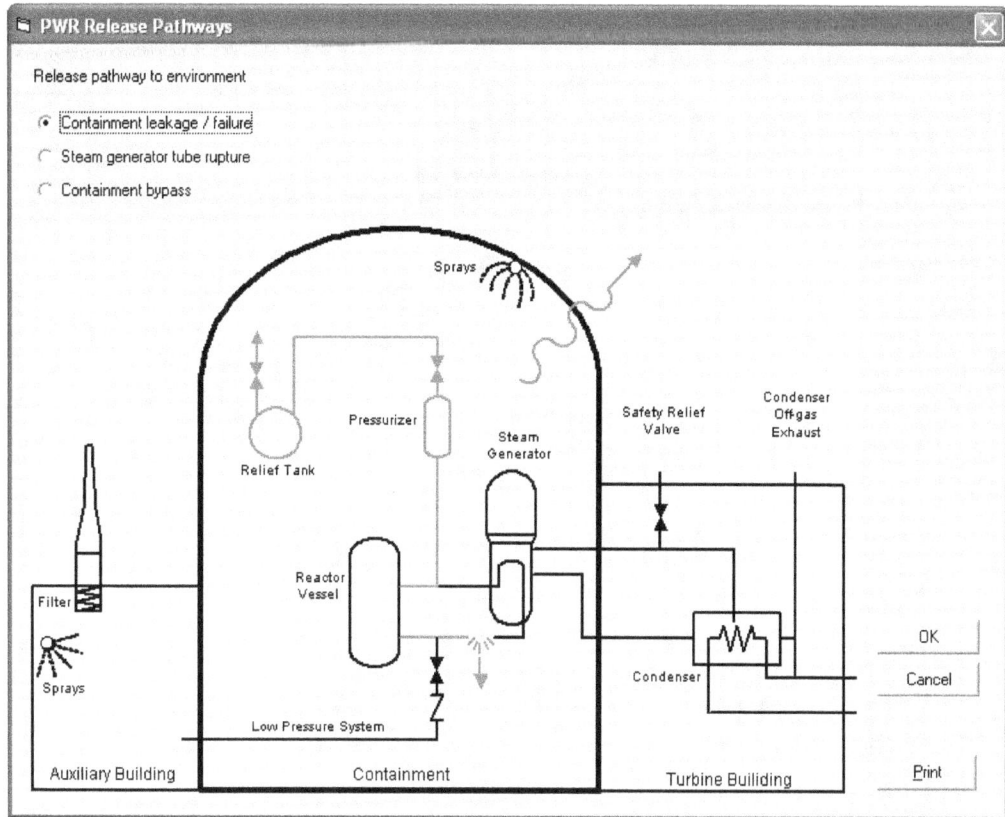

Select the option **Containment leakage / failure** and then click **OK** to display the containment release pathway definition screen.

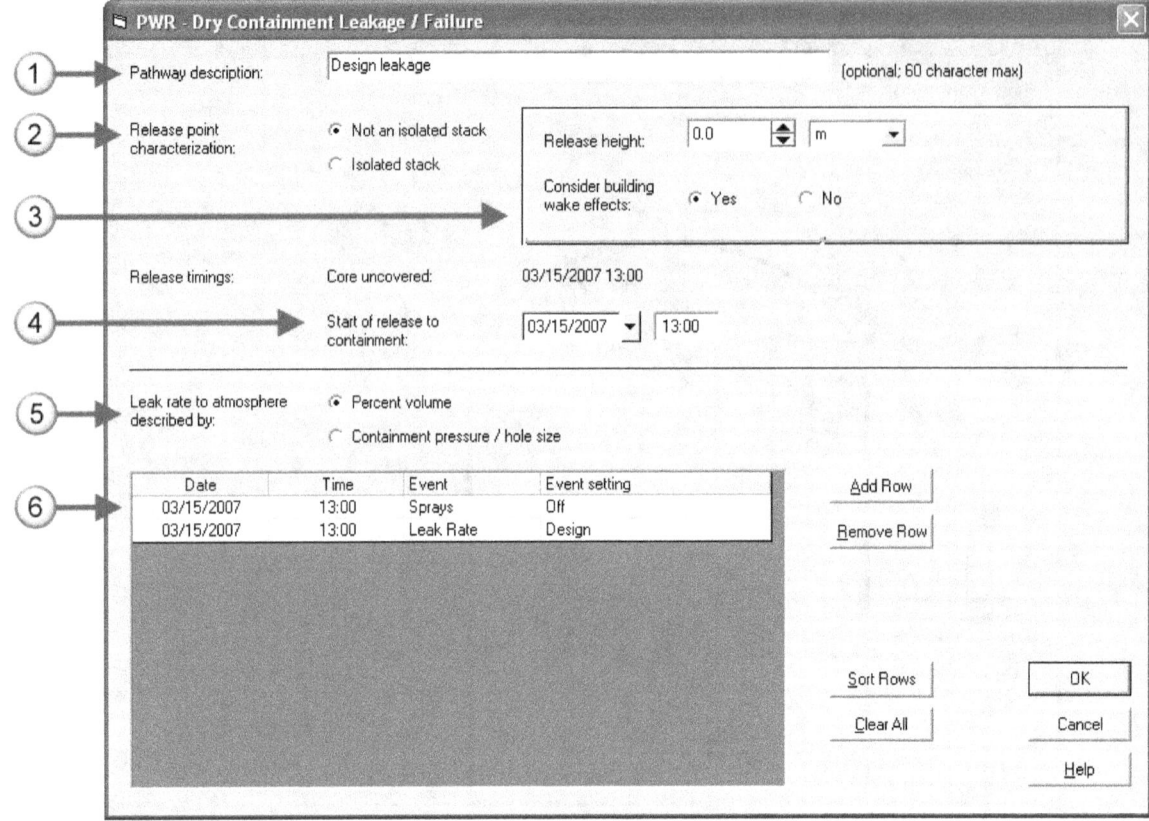

1. For the pathway description enter the text: **Design leakage**.

   This user-entered description is not used in the calculations but will appear on the case summary output. It should be used to clarify where or how the material is entering the atmosphere.

2. Leave the release point characterization set to: **Not an isolated stack.**

   This will be a release from the reactor containment building and not from a stack. Only some BWRs actually have isolated stacks.

3. Release height: **0.0 m**

   If known, the height above ground level at which the material enters the atmosphere should be entered. This value may be adjusted upward if needed to represent an effective release height. For example, if a fire is forcing the material higher you might use an estimate of the height of the smoke as an effective release height. In our case we do not have a defined release point so leave the release height at the default value of **0.0 m**.

   Consider building wake effects: **Yes**.

   Historically, adjustments to the atmospheric dispersion parameters have been made to compensate for releases from buildings. These adjustments have been based on building dimensions and have not considered wind speed. RASCAL includes optional adjustments to the dispersion parameters that are functions of building area, wind speed, and distance downwind (Ramsdell 1990). The adjustments, which are inversely related to wind speed, start at zero at the release point and increase with distance downwind until a limit, based on building area, is reached. (The adjustments are actually a function of travel time.) When the wind speed exceeds about 4 m/s the adjustments become negligible. The adjustments have been referred to as building wake corrections. However, the adjustments are more sensitive to wind speed than building area and are only significant at low wind speeds. Therefore, it is more appropriate to refer to the correction as a ***low wind speed correction***. More recent analysis of the experimental dispersion data (Ramsdell and Fosmire 1998a, 1998b) has yielded adjustments that are independent of building area.

4. Leave the **Start of release to containment** time as initially set to match the core damage time (**13:00**).

   This entry allows for the consideration of the holdup of material before it enters the containment atmosphere. No information is provided on any delays in the radioactivity moving from the core to the containment. Thus, assume that the material enters the containment immediately.

5. Leak rate to atmosphere described by: **Percent volume**

   Leakage fraction for reactors can be described in two ways: 1) as a leak rate from containment in percent volume per time, or 2) as a leak rate set by a pressure and containment hole size. Both methods allow the leak rate to vary with time. The methods cannot be combined.

6. Leave the spray and leak rate events unchanged

   The lower portion of the screen deals with describing the time dependent leakage of material to the environment. A leak type is selected and then release events are entered into a table. The default is for a leak rate type of percent volume, for there to be no active reductions (e.g. sprays, filters), and for the leak rate to be at design. The event grid already has 2 events defined: sprays off and design leak rate. The defaults are correct for the problem and do not need to be changed.

Click the **OK** button to accept the release pathway definitions.

In many of the later problems in the workbook the inputs are described in a much briefer format without all the discussion. However, when appropriate, the rationale for an input will be provided.

Following is the abbreviated input summary for the problem up to this point.

Inputs

Event Type	Nuclear Power Plant
Event Location	Arkansas - Unit 1
Source Term	Time Core is Uncovered
	Reactor shutdown: **10:00**
	Core uncovered: **Yes, at 13:00**

Core recovered: **No**

Release Pathway

Containment Leakage / Failure

Pathway description: **Design leakage from containment**

Release point characterization: **Not an isolated stack**

Release height: **0.0 m**

Consider building wake effects: **Yes**

Start of release to containment: **13:00**

Leak rate to atmosphere described by: **Percent volume**

Release path events:

\<date\>	**13:00**	**Sprays**	**Off**
\<date\>	**13:00**	**Leak Rate**	**Design**

5 Entering Site Meteorological Data

Purpose

To learn how to enter the basic meteorological observations obtained from the site.

Discussion

Meteorological data is needed to model the behavior of the radioactive plume in the atmosphere. The minimum required data is:

wind speed and direction	defines the downwind transport
atmospheric stability	governs plume horizontal and vertical growth
precipitation	defines wet deposition (a very important removal process)

The first step in creating a meteorological data set for use with RASCAL is entering the meteorological observations for the release point. The data can be obtained from a variety of sources. For nuclear power plants, observations for the site itself would be obtained via the Emergency Response Data System (ERDS) or by telephone from the licensee. We will start by entering that data.

ERDS has limitations:

- it provides no forecast meteorological information
- it may not be connected when needed
- the data is from a single location and may not be sufficient to describe winds in complex terrain
- there are no ERDS connections at facilities other than nuclear power plants

Problem

For purposes of this problem, we will assume that ERDS and the plant operators have provided the following data:

Time	Wind direction (deg)	Wind speed (mph)	Stability class	Precipitation	Air Temp (deg F)
1100	10	6	B	No precip	65

Note that ERDS does *not* provide information on precipitation. That information would need to be acquired from another source.

Select the **Meteorology** button on the main screen. This will display the **Select Meteorological Data Set** screen.

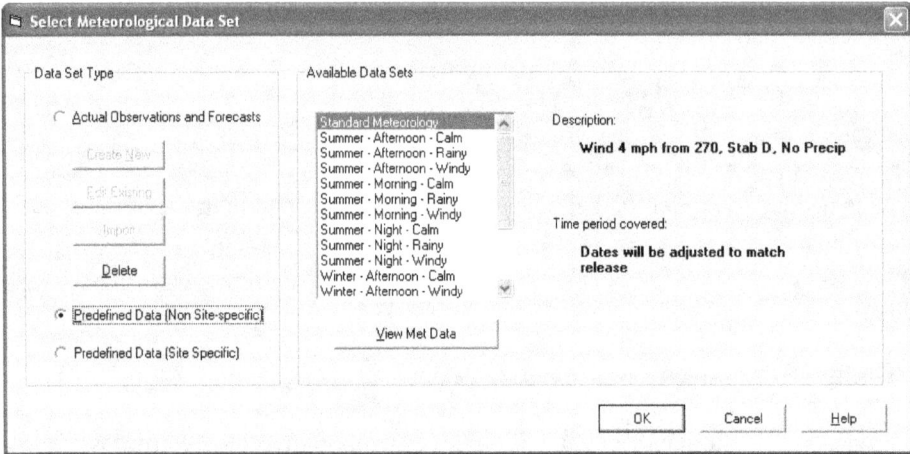

The default is to use **Predefined Data (Non Site-specific)**. The available data sets portion of the screen lists all the data distributed with the RASCAL program. These "generic" data sets are described further in Topic 40.

Actual Observations and Forecasts

This data type is both site-specific and date/time-specific. That is, it can be used only for releases from the site specified at the specified times. Since it is site-specific, it can correct for topographic effects. The date and times of the meteorological data must closely match the release times. Meteorological data must start within a 2 hour window before the release to the atmosphere starts. The data set can be comprised of a mix of observations and forecast records. However, entering a set of observations for a station will cause any forecasts for that station at an earlier time to be deleted.

1. Choose the **Actual Observations and Forecasts** data set type.

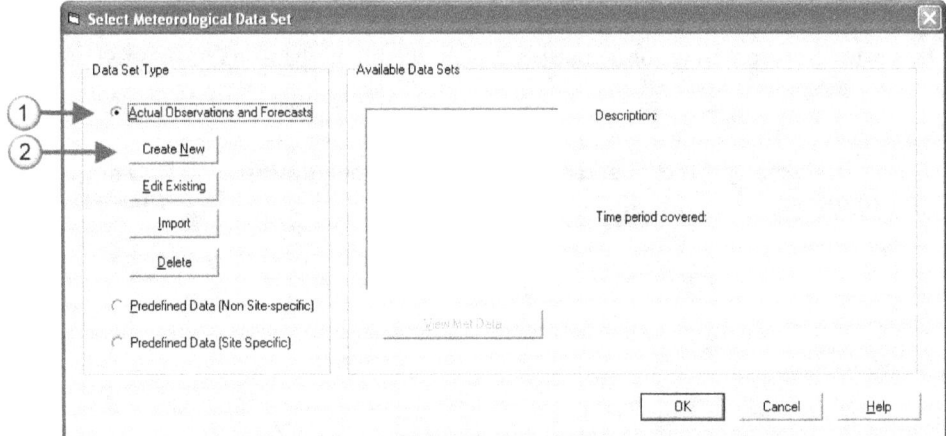

This will enable the buttons used to access the **Meteorological Data Processor** program. If no data sets have been previously created for this site, then the **Available Data Sets** box will be empty.

2. Click the **Create New** button.

Since we have not yet entered any weather data for this event, we want to create a new set. Using actual data requires that the times of the meteorological observations closely agree with the times of the release to the atmosphere.

This will start the **Meteorological Data Processor** program and display the main screen.

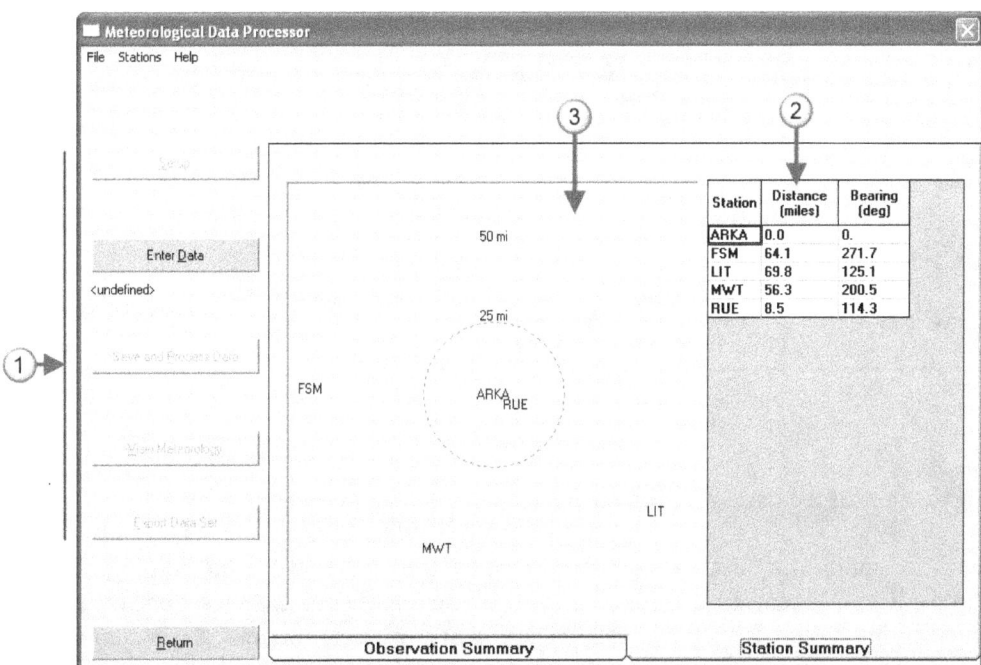

1. The main steps of the program are controlled via the buttons on the left side; starting from the top and working downward. The **Location** button is greyed out and unavailable because the program was started from within the source term to dose model. At this point the location has already been defined and thus doing it again is not necessary. The **Location** button is available only if the **Meteorological Data Processor** program is started in a stand-alone mode. The **Save and Process Data**, **View Meteorology**, and **Export Data Set** buttons are also unavailable since no data has yet been entered.

2. This is a list of the available weather stations surrounding the site. These are taken from the RASCAL database. The display shows the distance and the direction of the station from the release point.

The NWS is continuing to add observing stations. If you find other stations that you would like to use they can be added. Refer to "Adding Weather Stations" on page 163 for the steps.

3. A simple map show the relative positions of the stations to the release point.

Click the **Enter Data** button to display the data entry screen.

Data Entry Screen

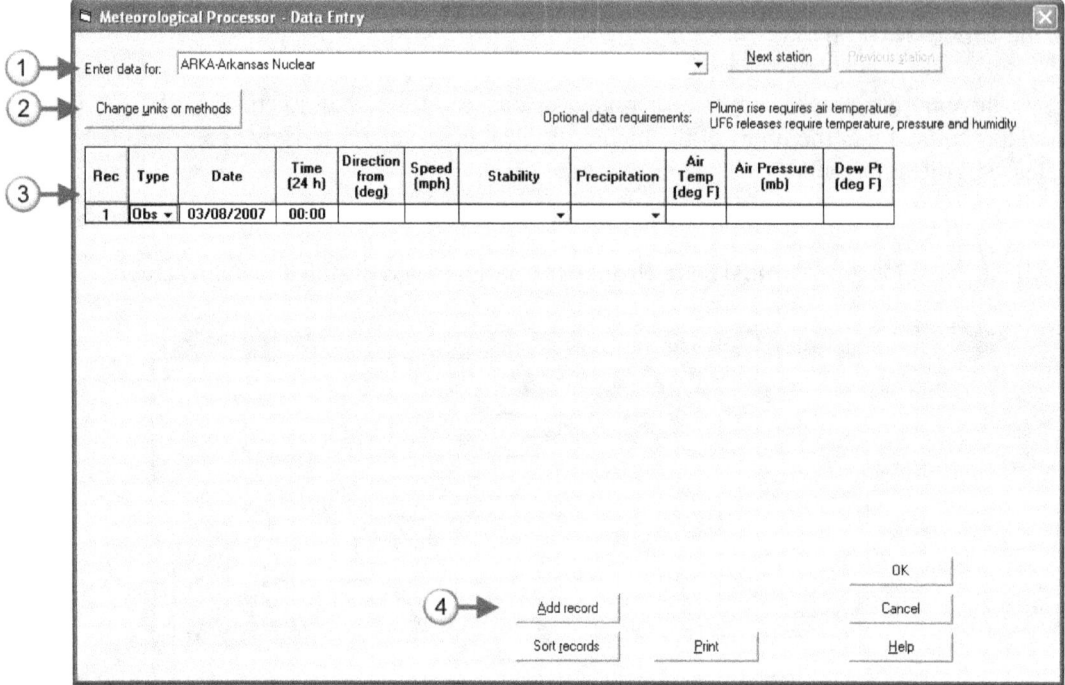

Note the following features of the data entry screen:

1. The default weather station (first in the drop-down list at the top) is the on-site meteorological tower. For sites not in the RASCAL database, this station will be the release point. At least one observation must be entered for this station. It should usually be at the first observation time. It is recommended that you always begin by entering the data for the release point before entering data for other stations.

    You can change the station by picking from the **Enter data for** list. Try this to see how it works. Notice that the table of data will change as the station changes.

2. Use the **Change units or methods** button if the units of the data you are receiving do not match those currently displayed in the data entry grid.

3. Weather data for each station is entered into a separate grid. Do not combine data from multiple stations into one table. The table of data initially consists of a single, mostly blank record.

    Data can be entered for any date or time. However, there must be data before or at the same time as the start of the release to the atmosphere.

4. The **Add record** button is used to add records to the end of the data entry grid.

Entering Data

Now, enter the following data for station ARKA:

Column header	User entry	Comments
Type	Obs	This means that the record is an *observation* and represents an actual set of measured values (in contrast to forecast)
Date	today	Use the default, which if the computer clock is correct, should be today.
Time	11:00	Enter the time using a 24-hour clock. The program will round the time to the nearest quarter-hour. For example, an entry of 12:07 would be changed to 12:00. Similarly, an entry of 14:22 would be changed to 14:30.
Direction from (deg)	10	This is the direction from which the wind is blowing. A wind from the north is 0 degrees, from the east is 90 degrees, from the south is 180 degrees, and from the west is 270 degrees.
Speed (mph)	6	
Stability	B	
Precipitation	No precip	
Air Temp (deg F)	65	
Air Pressure (mb)	<blank>	This is needed only if the release contains UF_6.
Dew Pt (deg F)	<blank>	This is needed only if the release contains UF_6.

The record will look like this:

Rec	Type	Date	Time (24 h)	Direction from (deg)	Speed (mph)	Stability	Precipitation	Air Temp (deg F)	Air Pressure (mb)	Dew Pt (deg F)
1	Obs ▼	03/15/2007	11:00	10	6	B ▼	No Precip ▼	65		

Now click the **OK** button to return to the main form.

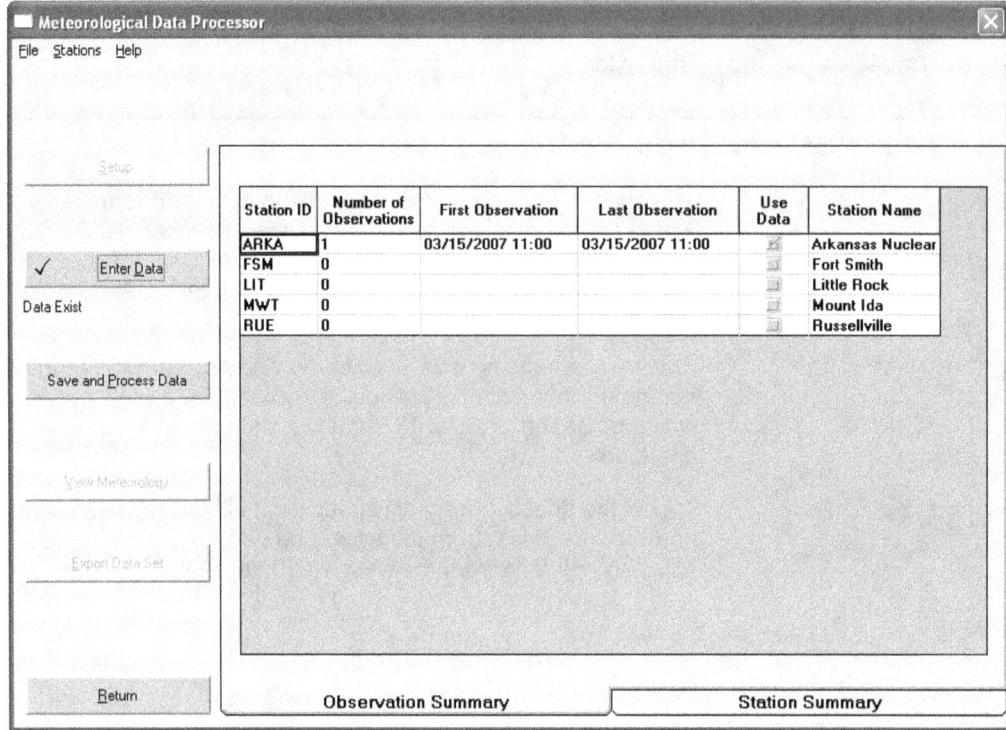

Observation Summary

Note that the **Observation Summary** tab is active when the screen is redisplayed after doing data entry. This summary shows in one place how much data has been entered. For each station, it shows the number of observations (or forecasts) and the date and time of the first and last entries. In this case, a single observation has been entered for the ARKA site.

Save and Process Data

The RASCAL transport and diffusion models require gridded wind fields for their calculations. Just entering the observation data into the program is not enough. A final step is required to take the information at these observation points and generate the gridded data. Each time additions or changes are made to the data, the save and process step must be repeated.

The **Save and Process Data** button is now active. Click it to display the **Save observations and process data** screen.

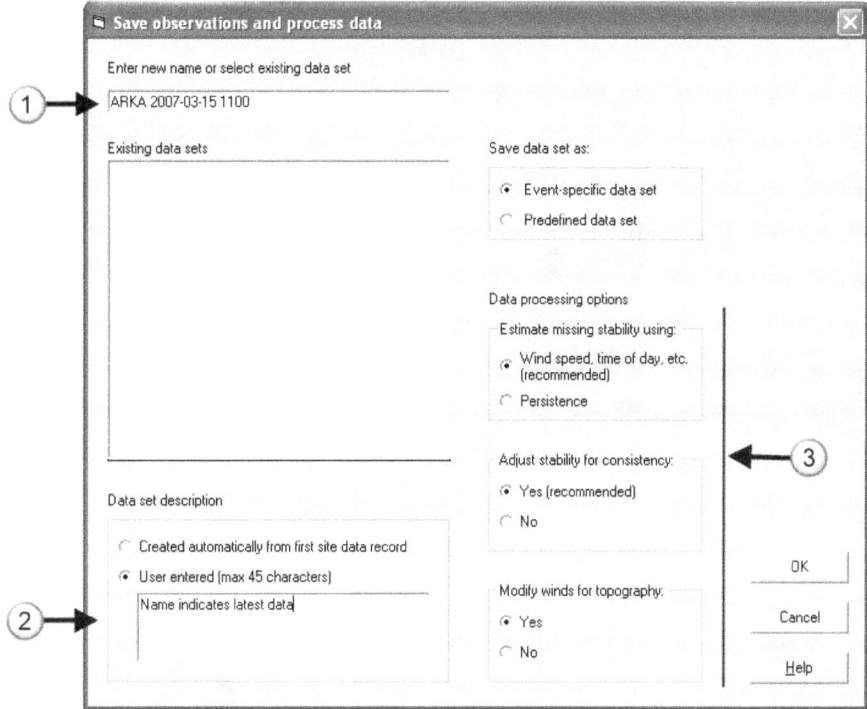

1. The field at the top of the screen is used to name the data set. This name should uniquely identify the data set to you and to others who will share the data set. It should consist only of alphanumeric characters. Punctuation characters (e.g. colons, commas, semi-colons) should be avoided.

   For this first data set, use a name that indicates when the data set was created, for example:

   ARKA 2007-03-15 1100   (Change the date in your name as needed)

2. The data set description is text that is displayed with the data set name in the data set selection screen. It is recommended that you *not* use the automatically created description. It will not change if the first data record stays that same (which is likely) and can be confusing.

   Instead, select the **User entered** option and then type in the reminder text "Name indicates latest data". This will remind you that the naming scheme uses the time the data was entered and the latest time should be the most current set of data.

3. Leave the **Data processing options** at the default settings

   - **Estimate missing stability using**

     This option allows the program to replace missing stabilities with an estimated value based upon wind speed, time of day, and precipitation. If **Persistence** is selected, the stability will remain the same as the previous observation.

   - **Adjust stability for consistency**

     This option allows the program to compare the user entered stabilities with ranges of stability classes that would be expected given the time of day and meteorological conditions. Stability values that are out of range are replaced with more likely values.

- **Modify winds for topography**

  The RASCAL database contains topography data for each site. This data can be used by the program to modify the wind fields so wind flows around mountains instead of through them.

Click the **OK** button. When the calculations are completed you will be returned to the main screen on the **Observation Summary** tab.

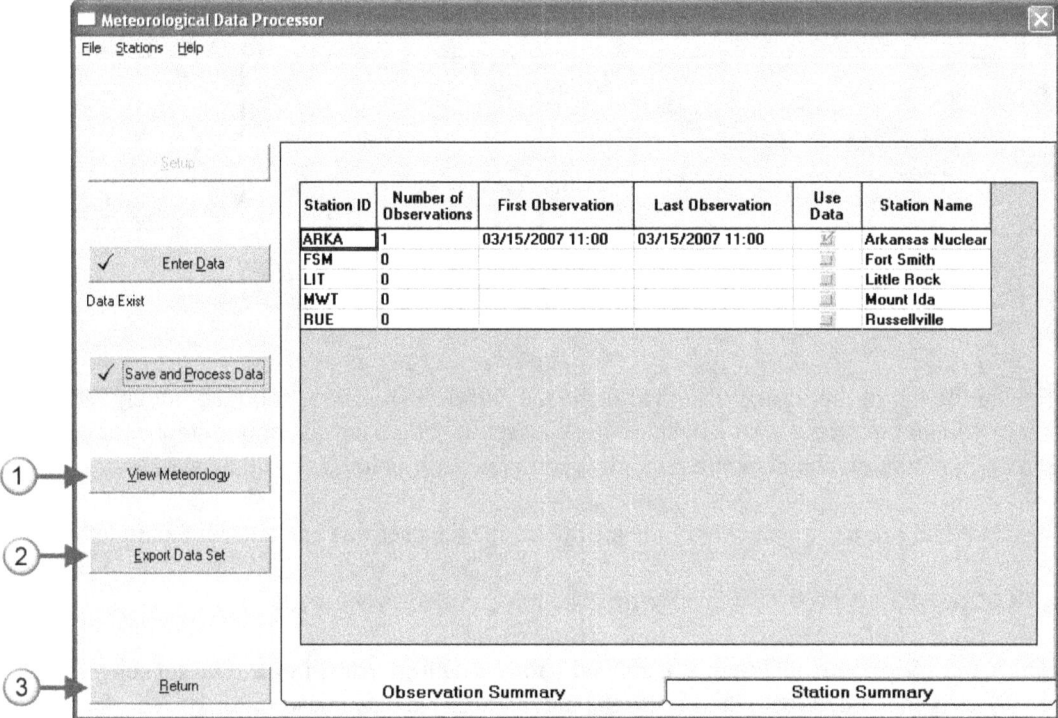

1. The **View Meteorology** button will launch a separate program for viewing both the entered data and the generated gridded data to be used by the models.

2. The **Export Data Set** button can be used to created a single file containing all the meteorological information in the current data set. This file can be saved in any location and can be copied or mailed. There is an import button on the meteorological data selection screen that allows this type file to be imported. Then export/import functionality allows the jobs of dose analyst and weather data collection and entry to be separated.

3. The **Return** button exits the Meteorological Data Processor and returns control to the Source Term to Dose model user interface.

Before returning to the model to use this data set, click the **View Meteorology** button.

Viewing the Meteorology

This is the **Meteorological Data Viewer** main screen.

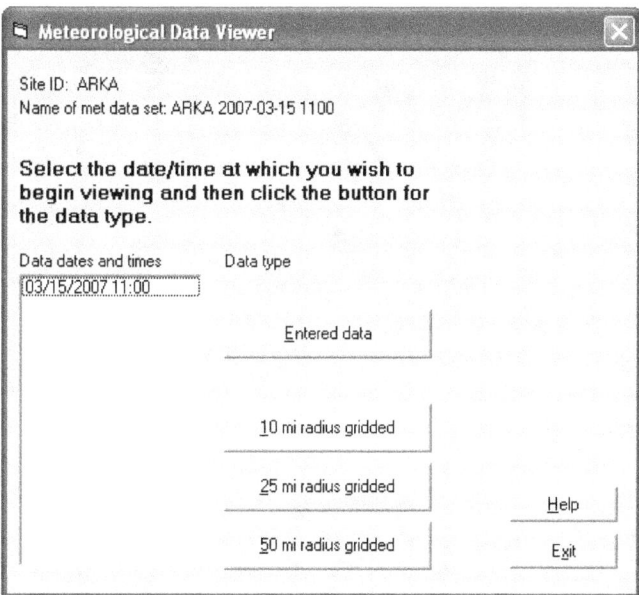

The viewer program provides both text and graphics displays of both the entered data and the processed data.

Click the **Entered Data** button.

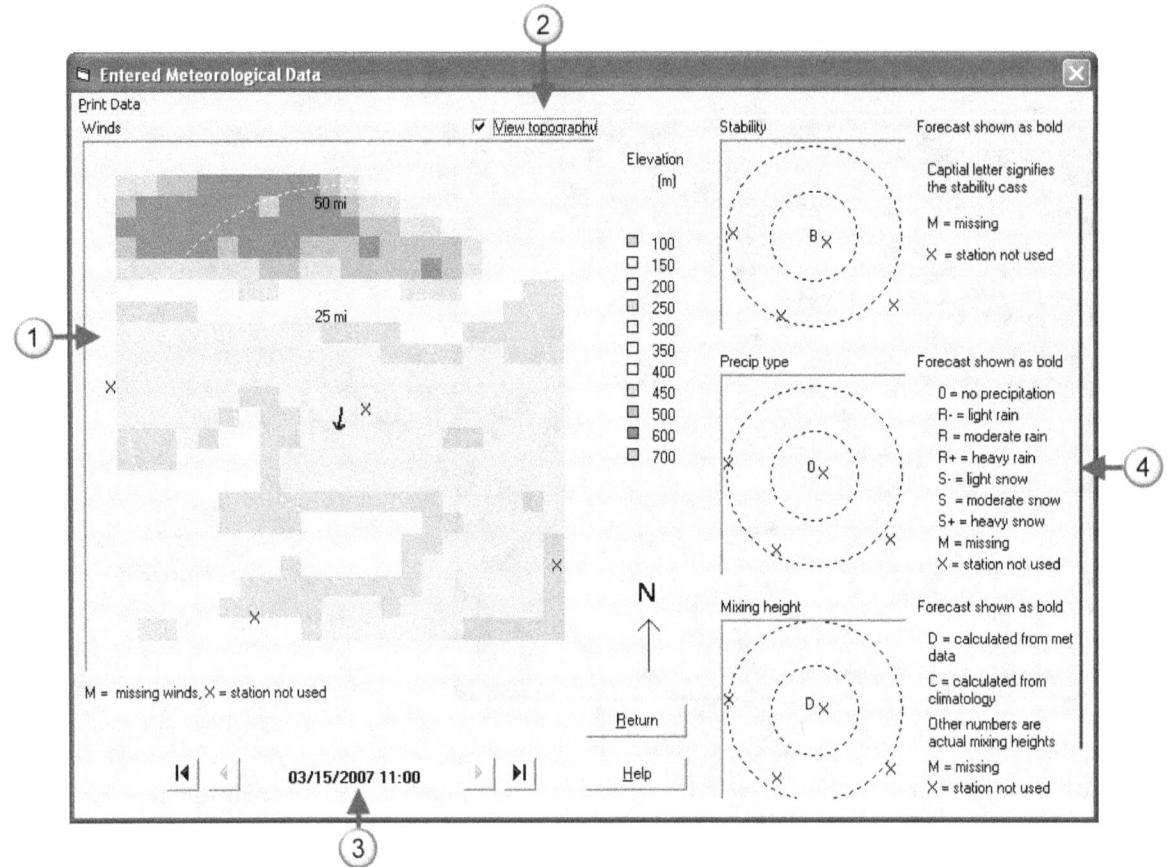

This screen provides a set of pictures to summarize the entered data at each time.

1. The large image on the left shows the winds. Arrows represent the direction in which the wind is blowing. In our case, the wind direction was 10 degrees (from the north). The length of the arrow is proportional to the wind speed. Since we entered only data for the site, there is no other station data to display and they are represented by an X.

2. The **View topography** check box controls whether a colored representation of terrain heights is displayed beneath the wind arrows.

3. The viewer can cycle through the data in 15 minute time steps. Since we had only a single measurement time there is only one time to display.

4. The 3 smaller images on the right reflect the data entered for stability, precipitation, and mixing height.

Click **Return** to go back to the data selection screen.

Click the **10 mi radius gridded** button.

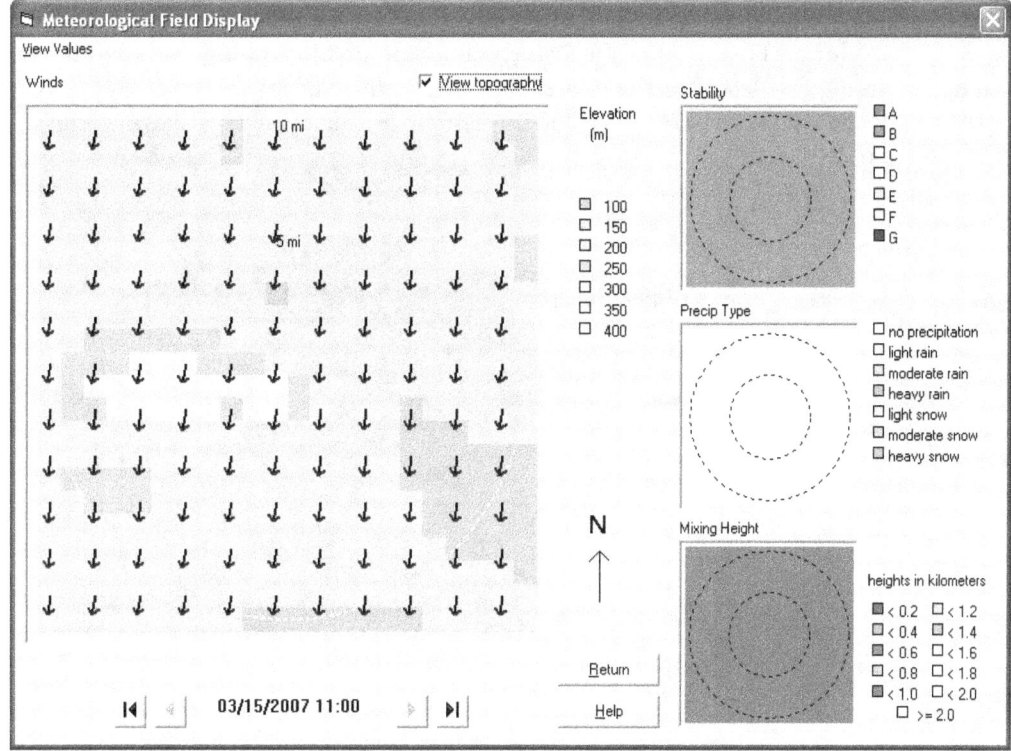

Now displayed in the large box on the left are arrows representing the wind field generated by the program from the single data record entered. Recall that the wind direction was 10 degrees. Wind direction is given as the direction *from* which the wind is blowing. Thus, the arrows show the direction material would be transported.

The three smaller boxes to the right show the stability, precipitation type, and mixing height. The first two are as entered with the data. The mixing height was calculated by the program.

We now have a group of files that include the user entered data and the processed wind fields to be used by the transport and diffusion models. This data set will work only with the Arkansas site and on the date specified. This data set can be recalled at any time for editing. However, each time the data is changed, you must again **Save and Process Data**.

Click the **Return** button to close the wind field display. Then, click the **Exit** button to close the Meteorological Data Viewer. Finally, click the **Return** button of the Meteorological Data Processor. This returns to the Meteorological Data Selection screen. Notice that the data set just created is selected for use. Also, notice that the data set description entered when processing the data is shown along with the date and time of the observation.

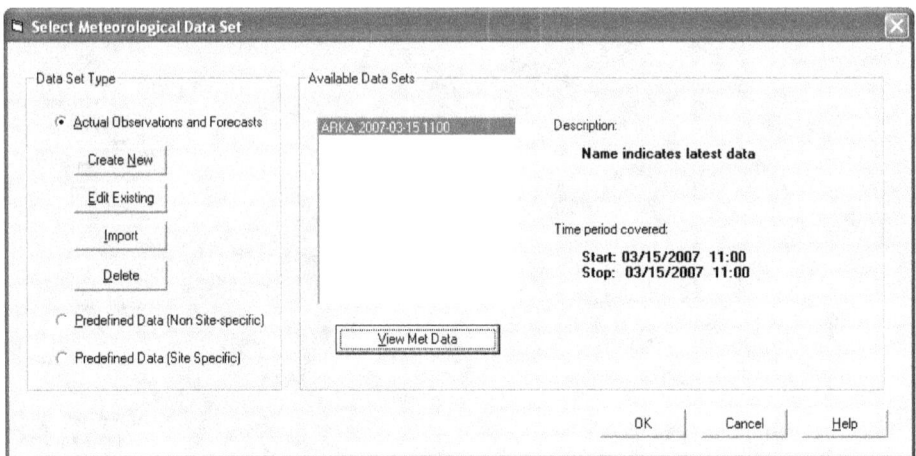

Click the **OK** button to return to the main screen.

6 Calculate Doses

Purpose

To learn how to calculate dose.

Discussion

The last information to be entered before starting the calculations is the distance to which doses are to be calculated and when to stop the calculations.

Setting the **Distance of calculation** determines which models are run and the locations for the model receptors. In addition to defining the distance to which doses should be calculated, you also need to specify when the calculations should be terminated. This is essentially the end-of-exposure. Doses are accumulated up to this time (with the exception of 4-day groundshine which always goes to 4 days). Setting this value is discussed more fully in Problem 11.

Problem

Recall that you have been asked to evaluate the appropriateness of the protective action recommendation. In order to do this you will need to determine the projected total effective dose equivalent (TEDE) and the thyroid dose (thyroid committed dose equivalent or thyroid CDE).

Inputs

Calculation Options

Distance of calculation: **Close-in + out to 10 miles (16 km)**

End calculations at: **Start of release to atmosphere plus 6 hours**
(We will look at this 6-hour assumption in greater detail later on)

Case description: **Core uncovered, design leak rate**

Click the **Calculate Doses** button on the main screen.

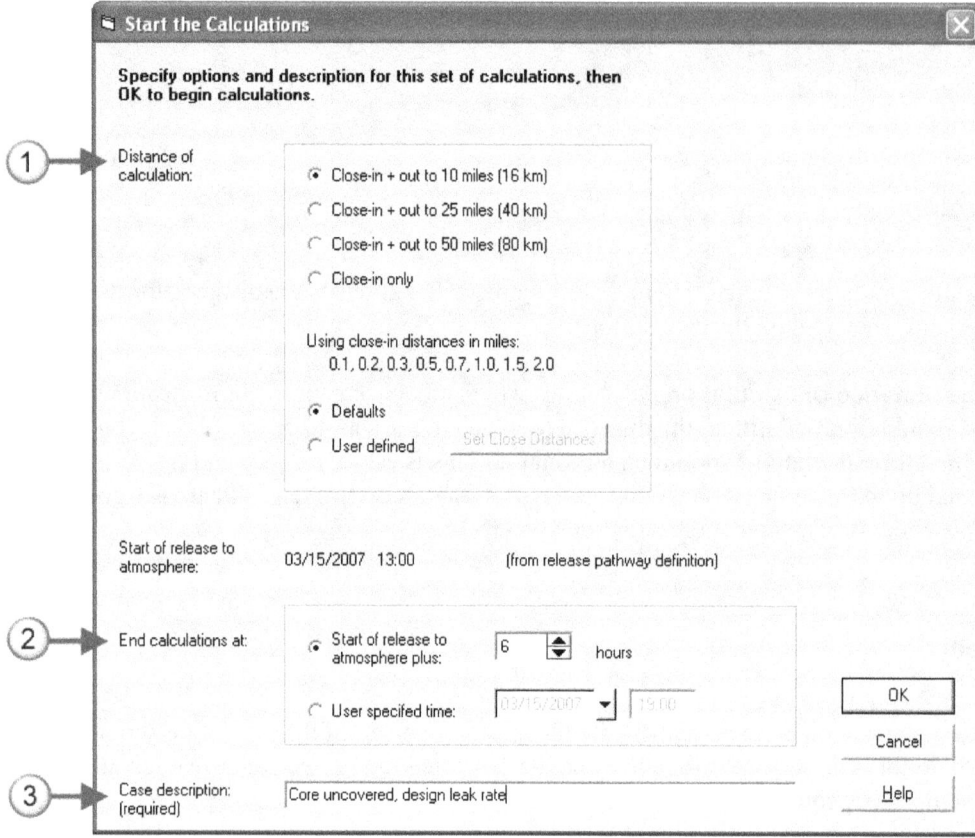

1. Distance of calculation: **Close-in + out to 10 miles(16 km)**

    Always select close-in plus out to 10 miles for the initial run. If there is a subsequent interest in doses at distances beyond 10 miles, run the calculations again selecting a greater distance. (Hint: if you save the case after you run it the first time, you will not have to re-enter all the data. You will just have to change a single entry on the **Start the Calculations** screen.)

2. End of calculations: **Start of release to atmosphere plus 6 hours**

    The default of 6 hours is sufficiently long that, even if the wind speed was low (e.g. 2 mph), some material would reach the default distance setting of 10 miles. You should always evaluate whether this "end-of-calculation" provides enough time given the wind speed and final distance of interest for the material transport. This issue is discussed in more detail in Problem 11.

3. Case description: **Core uncovered, design leak rate**

    This text will be shown on all output products and should uniquely describe the case you are running.

Click the **OK** button to start the calculations.

Status windows will indicate the progress of calculations. The sequences of calculations is: source term, close-in model, and if needed the puff model to the specified distance. (The transport and diffusion models are described in Topic 7). At the end of the calculations, the Maximum Dose Values tab of the main screen will be displayed.

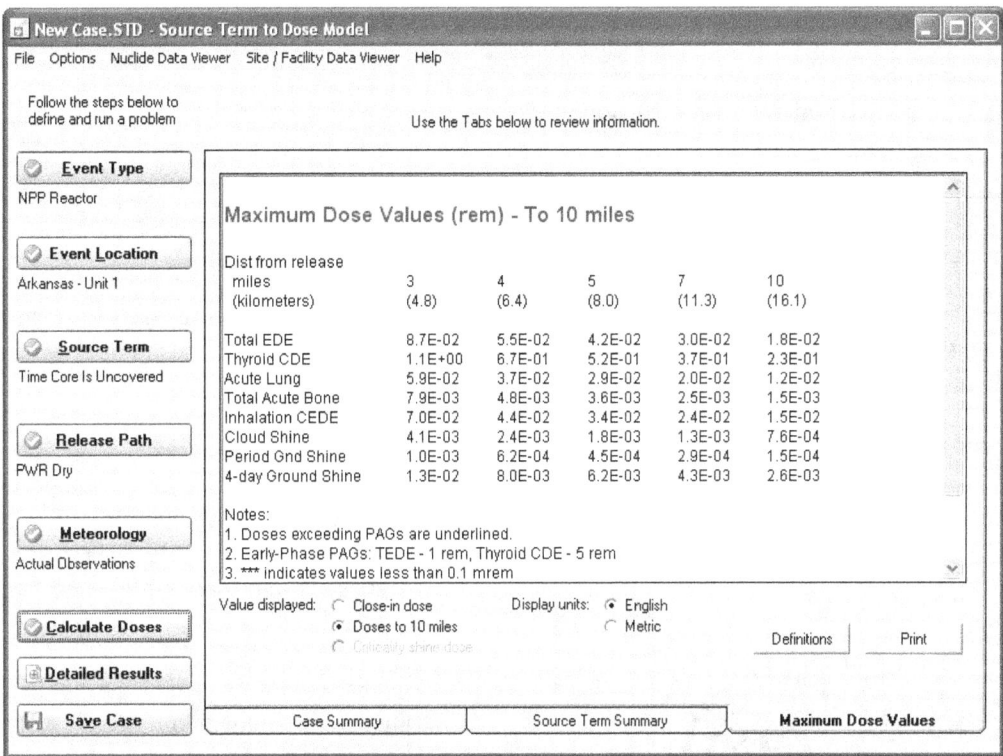

SAVE THIS CASE!!

This is the time to save all your work so that it can be recalled later. Click the **Save Case** button in the lower, left of the main screen or select **File | Save Case** from the menu. Enter a name for the file and click **OK**.

For consistency in the workbook, use the name: **Using RASCAL - Case 1**

All the input information and all the results will be saved in a single file that can be backed-up, copied, and distributed.

7 Transport and Diffusion Models

Purpose

To learn about the two transport and diffusion models used in RASCAL.

Discussion

There are two transport and diffusion models in RASCAL used for radiological releases (a separate plume model handles UF_6 releases). Close to the release point a straight-line Gaussian plume model is used. This model is characterized by the following features:

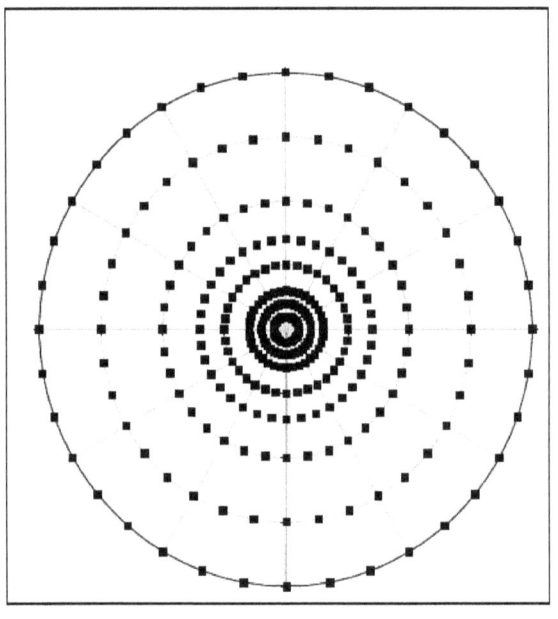

-   A polar grid of receptors is used. As shown in Figure 1, this grid consist of 8 concentric circles with receptors located every 10 degrees around the circles. There are default distance settings (see Table 3) for the circles or the 8 distances may be set by the user.

-   The model is steady-state. That is material is transported "immediately" to all distances. There is no need to allow for transport time on this grid. The disadvantage is that the receptors may report doses before the material has had time to be transported to the receptor.

-   The release point is always at the center and space between the receptors on a circle increases with distance from the release.

Figure 1 Polar receptor grid for the plume model

Table 3 Default distances for close-in polar grid

Calculation distance setting	Distances in miles							
Close-in only or Close-in + 10 mile	0.1	0.2	0.3	0.5	0.7	1.0	1.5	2.0
Close-in + 25 mile	0.25	0.5	1.0	1.5	2.0	3.0	4.0	5.0
Close-in + 50 mile	0.5	1.0	1.5	2.0	3.0	5.0	7.0	10.0

Further away from the release point, RASCAL switches to a Lagrangian puff model. This model has the following characteristics:

- It uses a uniformly spaced Cartesian grid of 41 x 41 receptor points. Figure 2 illustrates this grid. The release point is always at the center.

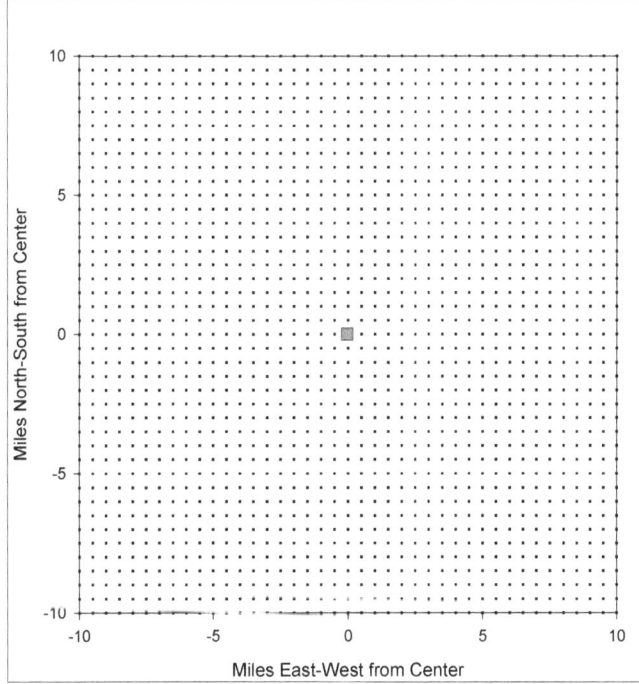

- The grid spacings are set by the user selecting from 3 different distances: 10, 25, or 50 miles. The resolution of the grid decreases as the calculation distance selected is increased. The distance between receptors for each calculation distance is shown Table 4.

- Time must be allowed for material transport time. For example, with a 4 mph wind speed, if the calculations are ended after only 2 hours, the material will not have had time to be transported to the 10 mile edge of the modeling area.

Figure 2 Cartesian receptor grid for the puff model

Table 4 Distances between Cartesian grid receptors

Calculation distance	Distance between receptor points
10 miles	0.50 miles
25 miles	1.25 miles
50 miles	2.50 miles

These receptor grids are evident when looking at dose footprints. With the closein polar grid, the footprint consists of wedge-shaped pieces representing the area around a receptor point as shown in the example below.

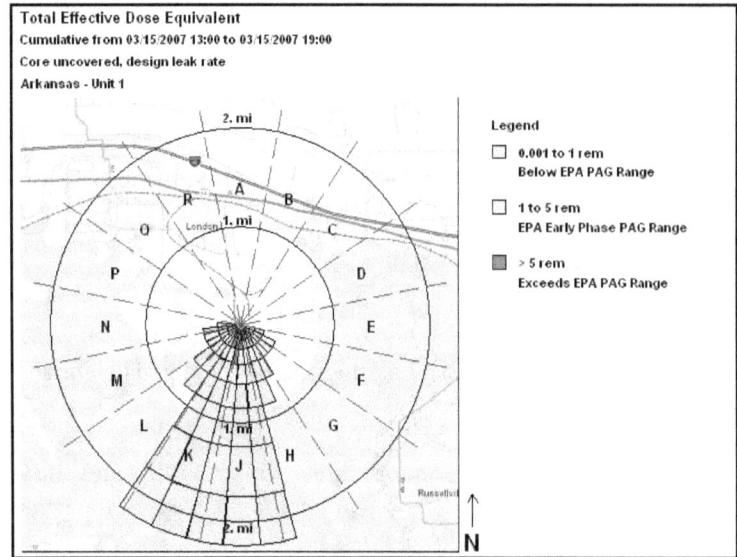

With the puff model, the footprint is represented by squares drawn around each receptor.

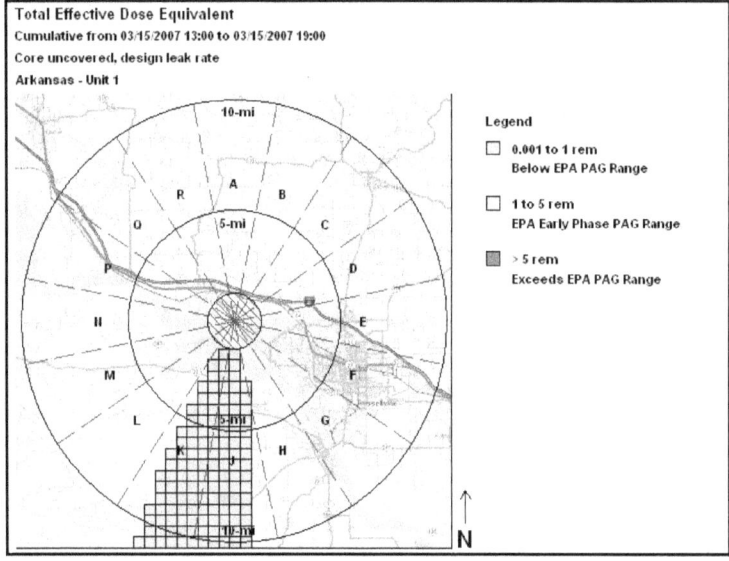

Remember:

The close-in calculation uses a straight-line Gaussian model. This is a steady-state calculation. That is, material is transported "immediately" to all distances. There is no consideration of the calculation duration. Receptors may report doses before the material has had time to be transported to the receptor. Use care when interpreting the results from the close-in model especially when using distances beyond 2 miles.

8 Using the Maximum Dose Values

Purpose

To learn how to use the RASCAL maximum dose values tables.

Discussion of PAGs

Chapters 2 and 5 of the EPA *Manual of Protective Action Guides and Protective Actions for Nuclear Incidents* (EPA 400-R-92-001) discuss the protective action guides (PAGs) for the early phase. The following material is from those chapters.

Protective Action	PAG (projected dose)	Comments
Evacuation (or sheltering)	TEDE 1-5 rem	Evacuation (or for some situations, sheltering) should normally be initiated at 1 rem TEDE or 5 rem thyroid CDE.
	Thyroid CDE 5-25 rem	
	TEDE is the sum of inhalation dose and the external doses from cloudshine and 4-days of groundshine.	

"Although the PAG is expressed as a range of 1-5 rem, it is emphasized that, under normal conditions, evacuation of members of the general population should be initiated for most incidents at a projected dose of 1 rem." (page 2-5)

"Sheltering may be preferable to evacuation as a protective action in some situations. Because of the higher risks associated with evacuation of some special groups in the population (e.g. those who are not readily mobile), sheltering may be the preferred alternative for such groups as a protective action at projected doses up to 5 rem. In addition, under unusually hazardous environmental conditions, use of sheltering at projected doses up to 5 rem to the general population (and up to 10 rem to special groups) may become justified." (page 2-5)

"Illustrative examples of situations or groups for which evacuation may not be appropriate at 1 rem include: a) the presence of severe weather, b) competing disasters, c) institutionalized persons who are not readily mobile, and d) local physical factors which impede evacuation." (page 2-6)

"No specific minimum level is established for initiation of sheltering. Sheltering in place is a low-cost, low-risk protective action that can provide protection ..." (page 2-7)

"The projected dose comparison to the early phase PAGs is normally calculated for exposure during the first four days following the projected (or actual) start of a release. The objective is to encompass the entire period of exposure to the plume and to deposited material prior to implementation of any further, longer-term protective action, such as relocation. Four days is chosen here as the duration of exposure to deposited materials during the early phase because, for planning purposes; it is a reasonable estimate of the time needed to make measurements, reach decisions, and prepare to implement relocation." (page 5-6)

Problem

Fill in the answers in the table below.

Question	Your answer
What are the TEDE's at 2 and 5 miles?	
What are the thyroid doses at 2 and 5 miles?	
To what distance does the TEDE exceed the PAG?	
To what distance does the thyroid dose exceed the PAG?	
At what time does the plume first reach 5 miles? (Hint: what was the wind speed?)	
At what time have you assumed the release to end? (Hint: On the "calculate doses" screen, look at the "end calculation at ..." entry)	
What was the time for the latest weather data that you entered?	

Optional question

RASCAL model results may be used to prepare briefings for decision makers. The information that goes into a briefing is discussed in more detail in Problem 16.

Should the results of this RASCAL run be used to prepare the briefing?	

Inputs

No new inputs are required.

Results

The first results screen displayed is the Maximum Dose Values table.

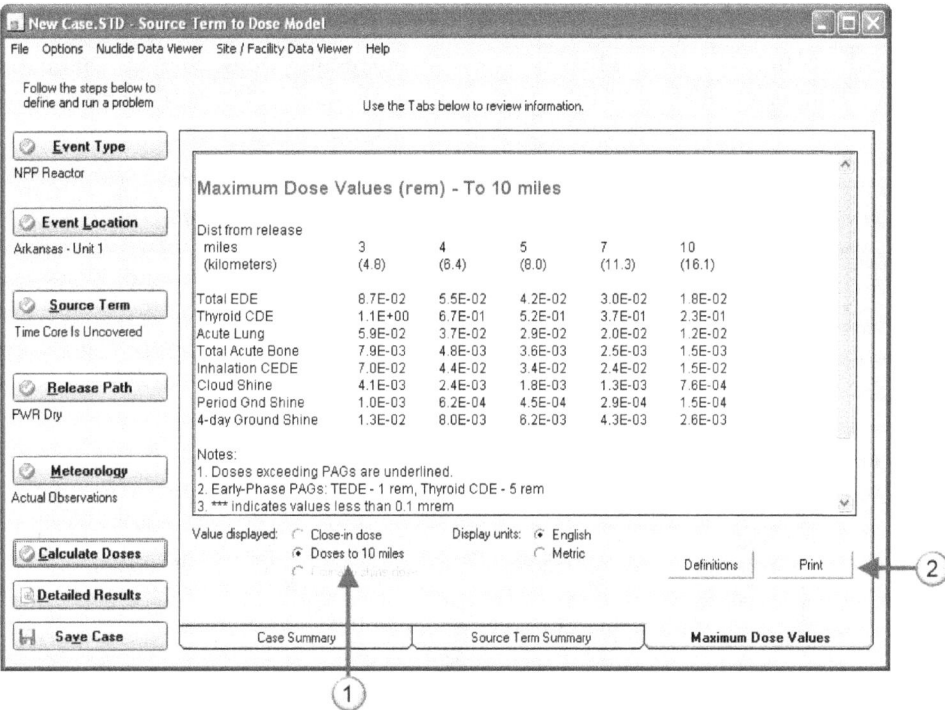

1. Use the **Value displayed** options to select which set of doses you want to see. The sample screen shown above is for the doses beyond the close-in area. A sample screen of close-in doses in shown on the next page.

    The doses shown are the highest doses at fixed distances from the release. The screen only shows at one time the "close-in" doses or the doses beyond the close-in area. The "close-in" doses are the doses calculated by the straight-line Gaussian plume. The doses beyond the close-in area are calculated using a Lagrangian puff model.

2. The tab contains a **Print** button to be used to send the dose table to the currently selected printer. Both sets of dose values are printed along with a complete description of all the data you have entered. Changing the printer is done using the **File | Print Setup ...** item on the menu bar.

The early phase (plume phase) doses that RASCAL calculates are for people who are outdoors during plume passage and who remain outdoors exposed to ground shine from deposited radionuclides for four days after the radionuclides have been deposited. Thus, the early phase doses that RASCAL calculates are larger than the doses that would be expected for people engaged in normal activities (spending much time indoors). In addition, people who evacuated or relocated within four days after plume passage would receive lower doses than RASCAL calculates.

This overestimation of the early phase doses by RASCAL is intentional. The purpose of the RASCAL dose calculation is to determine if protective actions are needed, not to provide realistic estimates of the doses that people would actually receive. The need for protective actions is based on the doses that would be received if no protective actions of any type were taken, even actions such as simply spending some time indoors. The RASCAL dose estimates should not be used as an estimate of the doses that would be received by people who did not intentionally take protective actions because even performing normal everyday activities will reduce doses to below those estimated by RASCAL.

The TEDE and Thyroid CDE numbers will appear underlined if the value exceeds the lower end of the EPA PAG range. Following the table of numbers is a "notes" section which defines the PAG thresholds and provides other information about the values shown. The maximum values table has a minimum dose of 0.1 mrem. Doses falling below that threshold are not shown. However, these low numbers can be examined by using the numeric values option for display under **Detailed Results**.

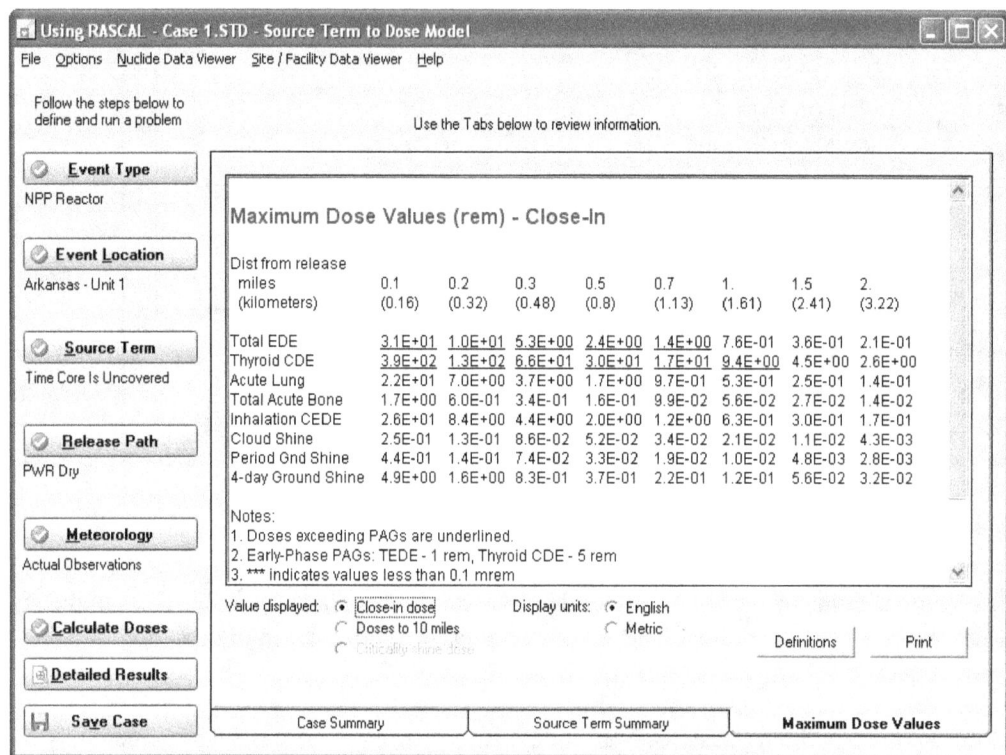

Use the **Close-in doses** and **Doses to 10 miles** views of the Maximum Dose Values screen to answer the first 4 questions that you were asked.

Did you get the answers below?

Question	Your answer
What are the TEDE's at 2 and 5 miles?	0.21 and 0.042 rem
What are the thyroid doses at 2 and 5 miles?	2.6 and 0.52 rem
To what distance does the TEDE exceed the PAG?	To between 0.7 and 1.0 miles
To what distance does the thyroid dose exceed the PAG?	To almost 1.5 miles
At what time does the plume first reach 5 miles?	About 2 P.M. (1hr after release start with 6 mph winds)
At what time have you assumed the release to end?	7 P.M. (Release start time of 1 p.m. plus 6 hours)
What was the time for the latest weather data that you entered?	11 A.M.

Answer to optional question

Should the results of this RASCAL run be used to prepare the briefing?	No. You can't use 11 A.M. weather to model plume behavior at 7 P.M. (8 hours later)

9 Obtaining Observed Weather Data from the Internet

Purpose

To learn how to obtain observed weather data from the Internet.

Discussion

Weather data from only the release point is not really good enough for dose projections to 10 miles.

As discussed earlier, site meteorological towers (e.g. those required at nuclear power plants) provide the best initial weather data for use by the RASCAL models. To supplement the on-site data and improve the representation of wind fields, we should use additional sources of weather data. The more data you have to represent the meteorological conditions in the modeling area, the better the wind field representation. Whenever possible, use data from multiple stations, especially stations downwind of the release point. This becomes especially important at greater distances (25 and 50 miles).

To obtain observed weather data:

- For sites in the RASCAL database, go to the website:

    weather.noaa.gov/weather/current/K???.html

    where you replace ??? with the three letter code for the weather station that you want. The station identifiers are case sensitive and must be all capital letters. The "html" must be lowercase. (The NWS website is picky).

    For example, the page with the current weather for Dulles airport would be **KIAD.html**.

- For sites not in the RASCAL database, go to the website: **weather.gov**

    Navigate to the area of interest using either the map or entering the city and state. The current weather observations at a nearby weather station are shown.

The National Weather Service is prepared to offer direct assistance if requested. They can provide observations as well as detailed forecasts. They will be able to give you additional insights on the data and tell you how reliable they think it is. In addition, they can provide stability class predictions and more exact information on precipitation.

The NWS requests that NRC staff contact the local forecast office (LFO) nearest the release. NRC Protective Measures Team procedures provide details on how to make contact and a list of offices and telephone numbers. The old procedure of first contacting the Senior Duty Meteorologist has been discontinued.

Non-NRC responders will need to establish their own relationship with the NWS.

Problem

Do the following if you have an Internet connection, otherwise skip down to the workbook data.

To supplement the weather data from the Arkansas site, obtain the current weather observations for RUE (Russellville) and MWT (Mount Ida). The Russellville station is 8.5 miles from the plant and will provide good additional data for computations to 10 miles. The Mount Ida station is 56 miles away but at least is in the downwind direction (currently).

If the wind speed units are not given, assume that they are in miles per hour (mph). You will find that the NWS does not provide stability on their web pages. You will need to get it from the NWS directly (call the LFO) or just enter "unknown" and let the code estimate it.

Fill in the table below with the current data obtained from the Internet:

Station	Type	Date	Time	Direction (from)	Speed	Stability	Precipitation
RUE	Obs						
MWT	Obs						

Workbook data to be used in working the problem

Normally, you would enter the information, as recorded above, into RASCAL. However, to avoid an inconsistency with the previous data, enter the data below.

Station	Type	Date	Time	Direction (from)	Speed (mph)	Stability	Precipitation
RUE	Obs	today	1100	340	5	unknown	No precip
MWT	Obs	today	1100	320	3	unknown	No precip

Begin, by clicking the **Meteorology** button to display the screen for selecting the data set.

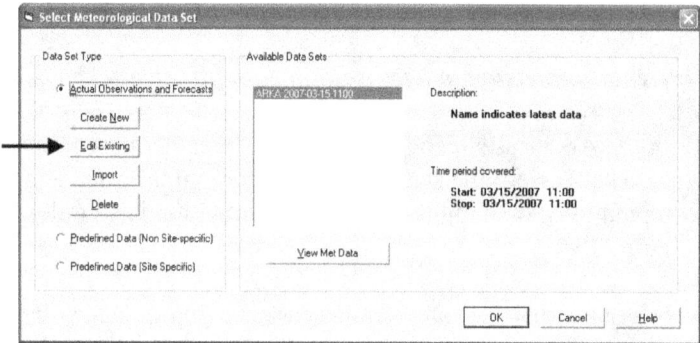

Click the **Edit Existing** button this time as we want to make additions to the data set not create a brand new one.

When the Meteorological Data Processor has started, select **Enter Data**. You will see that the single ARKA observation entered previously is still there. Now, we want to add the observations from the two NWS stations.

Remember that each station has its own data table. Select from the **Enter data for** choices to switch between the data tables.

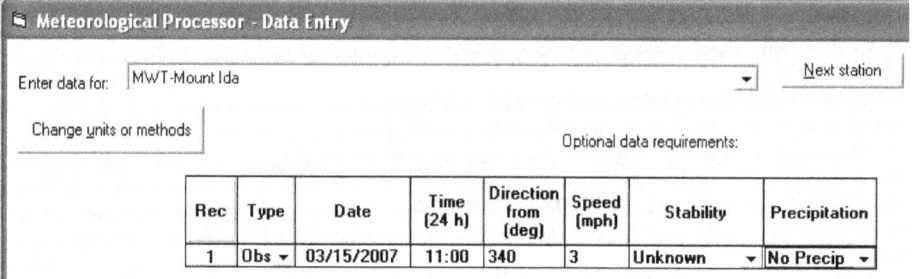

When both the RUE and MWT data has been entered, click the **OK** button. The **Observation Summary** tab now shows that data for 3 locations has been entered.

Station ID	Number of Observations	First Observation	Last Observation	Use Data	Station Name
ARKA	1	03/15/2007 11:00	03/15/2007 11:00	☑	Arkansas Nuclear
FSM	0			☐	Fort Smith
LIT	0			☑	Little Rock
MWT	1	03/15/2007 11:00	03/15/2007 11:00	☑	Mount Ida
RUE	1	03/15/2007 11:00	03/15/2007 11:00	☑	Russellville

Remember that the final step after changing any data is to process and save. Click the **Save and Process** button. Since we are editing the 1100 data set, we will get a warning asking whether we want to overwrite the existing file.

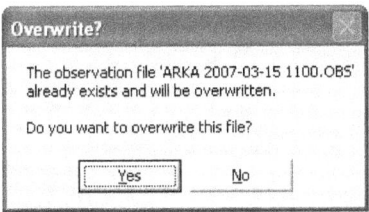

We do not want to overwrite the file so click the **No** button.

For this exercise, continue the naming scheme using the site ID and the date and time the data is updated. Assume that we are updating this data about half an hour later than before. Thus, use a name such as **ARKA 2007-03-15 1130.** By updating the time in the name we can always tell which data set was created last and presumably contains the "best" estimate of the conditions.

Recalculate the doses using this new data set to see if including more meteorological data makes a difference. Provide a new set of answers to the questions below.

Question	Previous answer (single station)	New Answer (3 stations)
What are the TEDE's at 2 and 5 miles?	0.21 and 0.042 rem	
What are the thyroid doses at 2 and 5 miles?	2.6 and 0.52 rem	
To what distance does the TEDE exceed the PAG?	To between 0.7 and 1.0 miles	
To what distance does the thyroid dose exceed the PAG?	To almost 1.5 miles	

SAVE THIS CASE!!

For consistency in the workbook, use the name: **Using RASCAL - Case 2.** You do not want to overwrite the previous case.

Question	Previous answer (single station)	New Answer (3 stations)
What are the TEDE's at 2 and 5 miles?	0.21 and 0.042 rem	0.21 and 0.049
What are the thyroid doses at 2 and 5 miles?	2.6 and 0.54 rem	2.6 and 0.61
To what distance does the TEDE exceed the PAG?	To between 0.7 and 1.0 miles	To between 0.7 and 1.0 miles
To what distance does the thyroid dose exceed the PAG?	To almost 1.5 miles	To almost 1.5 miles

There are no significant changes to either the doses or the distances.

10 Obtaining Forecast Weather Data from the Internet

Purpose

To learn how to obtain forecast weather data from the National Weather Service.

Discussion

Observed data is of course the best to use as input to the models. These are conditions that have actually been measured. This is fine if the release has already occurred and your job is to estimate where the material went. However, you may be doing an assessment of potential consequences at some future time. The release may not yet have taken place but you want to run scenarios based on possible future meteorology. Or, the release may be occurring now and you want to predict the movement of the plume for the next few hours.

Over short periods of time (a couple of hours) assuming the current observed conditions will persist may be adequate. For longer times, the uncertainty increases. Therefore, you should add forecasts to the meteorological data set.

Note that RASCAL will delete any forecast records that are followed by a later observation. Forecast records are only used for times later than the last observation.

To obtain forecast data,

1.   Go to the website: **weather.gov**

2.   Navigate to the site area by entering the city and state or by clicking on the map. Using the **Detailed Point Forecast** map, click as close to the site as possible.

3.   Select **Tabular Forecast** under the heading **Additional Forecasts & Information**.

4.   Record relevant information: time, wind direction, wind speed, probability of precipitation, and precipitation type and likelihood.

5.   If there is precipitation in the forecast, you will also have to go to the **Hourly Weather Graph** to get the amount of precipitation during each 6-hour interval.

Problem

Do the following if you have an internet connection, otherwise skip down to the workbook data.

Obtain the weather forecast for the Arkansas plant for 1 P.M., 4 P.M., and 7 P.M. Record the forecast data in the table below.

Hour	Wdir		Wspd	PoP	Precip type	Precip likelihood	Precip amount
	compass	degrees					
1300							
1600							
1900							

The National Weather Service gives wind direction in terms of compass directions (for example, N, NNE, NE, etc.) that the wind comes *from*. You will have to convert these to degrees.

N	= 000	E	= 090	S	= 180	W	= 270
NNE	= 023	ESE	= 112	SSW	= 202	WNW	= 292
NE	= 045	SE	= 135	SW	= 225	NW	= 315
ENE	= 067	SSE	= 158	WSW	= 248	NNW	= 338

Workbook data to be used in working the problem

For consistency, we will enter the data below into RASCAL. Modify the latest data set to add to the Arkansas data the three forecast records shown below:

Type	Date	Time	Direction	Speed	Stability	Precip	Air temp
Fcst	today	13:00	315	8 mph	B	None	68 F
Fcst	today	16:00	270	5 mph	D	None	76 F
Fcst	today	19:00	225	3 mph	E	None	75 F

Remember, you are *adding* 3 forecast records to the already existing observation for ARKA. On the data set selection screen, use the **Edit existing** button again with the **1130** data set selected. Then, in the meteorological data processor, click the **Enter data** button and add the forecast records.

The data grid for Arkansas will look like this:

Rec	Type	Date	Time (24 h)	Direction from (deg)	Speed (mph)	Stability	Precipitation	Air Temp (deg F)	Air Pressure (mb)	Dew Pt (deg F)
1	Obs	11/07/2006	11:00	10	6	B	No Precip	65		
2	Fcst	11/07/2006	13:00	315	8	B	No Precip	68		
3	Fcst	11/07/2006	16:00	270	5	D	No Precip	76		
4	Fcst ▾	11/07/2006	19:00	225	3	E	▾ No Precip ▾	75		

Process the data set and save with a new name (do not overwrite the previous file). In this case, use a time in the name of **1200**.

Use the **View Meteorology** button to start the viewer program.

Note that there are now 33 dates and times shown. The processor has filled in the gaps between the hours with additional data.

Click the **Entered data** button to examine the data.

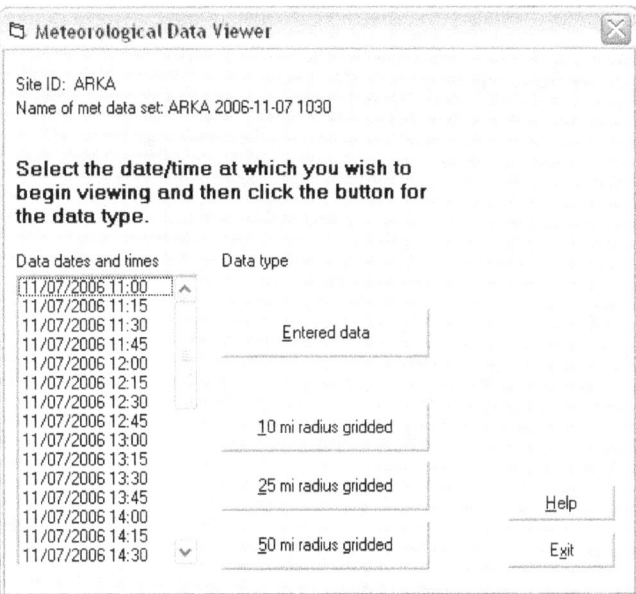

Step through the data using the control buttons and observe the wind arrow. Note that the arrow direction changes abruptly on the hour. This is because there is no interpolation between an observation and the forecast records.

Exit the Meteorological Data Viewer program and then return from the Meteorological Data Processor program into the **Select Meteorological Data Set** screen. The data set just created should be highlighted.

Run the calculations using this new data set to estimate doses using the following settings:

- Distance of calculation: Close-in + out to 10 miles
- End calculations at: Start of release to atmosphere plus: 6 hours

Answer the set of questions again based on your results using observed *and* forecast weather.

Question	Previous answer (3 stations - obs only)	New Answer (3 stations- obs & fcst)
What are the TEDE's at 2 and 5 miles?	0.21 and 0.045 rem	
What are the thyroid doses at 2 and 5 miles?	2.7 and 0.56 rem	
To what distance does the TEDE exceed the PAG?	To between 0.7 and 1.0 miles	
To what distance does the thyroid dose exceed the PAG?	To almost 1.5 miles	

Optional question

Should this new RASCAL run be used to prepare the briefing?	

SAVE THIS CASE!!

For consistency in the workbook, use the name: **Using RASCAL - Case 3**

Results

Question	Previous answer (3 stations - obs only)	New Answer (3 stations - obs & fcst)
What are the TEDE's at 2 and 5 miles?	0.21 and 0.045 rem	1.2 and 0.29 rem
What are the thyroid doses at 2 and 5 miles?	2.7 and 0.56 rem	14 and 3.4 rem
To what distance does the TEDE exceed the PAG?	To between 0.7 and 1.0 miles	~ 2.5 miles
To what distance does the thyroid dose exceed the PAG?	To almost 1.5 miles	To 4 miles

Adding the forecast information has resulted in increases in the projected doses. However, you are not seeing what direction the plume is moving.

Answer to optional question

Should this new RASCAL run be used to prepare the briefing?	No. You still do not have the information you need for your briefing. Continue with the next problem.

11 Time for End of Calculation

Purpose

To show how to select an appropriate time for the end of the calculations.

Discussion

On the start calculations screen you must specify the time after the start of the release at which you want to end the calculations.

The time can be specified as a number of hours after the start of the release to the atmosphere or as a specific date and time.

This is what happens at the time that the calculations end:

1. The release of material to the atmosphere ends (if it has not already ended).

2. The plume movement stops outside the close-in area.

3. Direct cloudshine dose from the plume stops.

4. Inhalation of radionuclides stops, but the full committed dose equivalent from radionuclides already inhaled is calculated.

5. Deposition of radionuclides on the ground stops, but the dose from 4-days of exposure to the radionuclides already deposited before the end of calculations is calculated.

Thus, if you end the calculations too early, projected doses may be too small because you have ended the calculations before the plume has delivered its dose. Conversely, if you end the calculations unnecessarily late, the computer run time will be unnecessarily long.

The diagram below illustrates the problem. Assume you have had a 1 hour release with a 4 mph wind speed. The ellipses represent the position of the plume in the air at various hours (H) after the release. The plume stretches for 4 miles. At H=1 the release has just ended. The leading edge of the plume has reached 4 miles (1 hr x 4 mph = 4 miles) and the trailing edge had just left the release point.

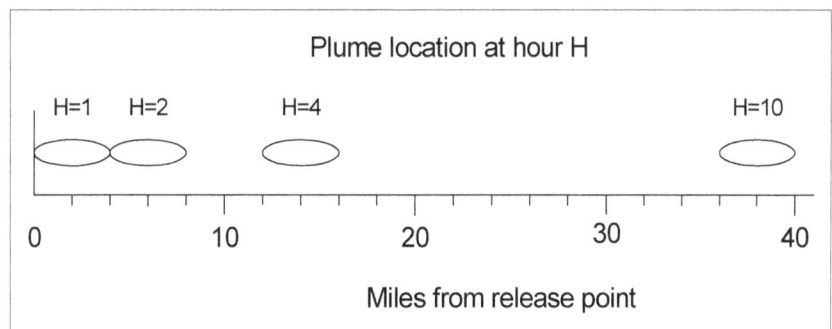

If you end the calculations 1 hour after the release starts, doses at 5 miles and beyond will all be zero. They will be zero not because no dose would be received but because insufficient time was allowed for material to be transported to those distances.

There is a simple rule of thumb that can be used for setting the calculation duration as long as the wind direction and speed and not going to vary too much over the model period.

$$Calculation\ Duration \geq Release\ Duration + \frac{Calculation\ Distance}{Wind\ Speed}$$

Round the calculation duration up to the nearest hour. However, be careful not to set a time much longer than necessary because it will increase the calculation time unnecessarily. Also, increase the duration if the wind directions are changing significantly and especially if they reverse directions. Remember not to mix incompatible units; use mph with miles or m/s with meters.

Remember:

As discussed in the earlier section on transport and diffusion models, the close-in calculation uses a straight-line Gaussian model. This is a steady-state calculation. That is, material is transported "immediately" to all distances. There is no consideration of the calculation duration. Receptors may report doses before the material has had time to be transported to the receptor. Use care when interpreting the results from the close-in model especially when using distances beyond 2 miles.

Problem

Assume that after 12 hours, the pressure in the containment will be reduced to zero. This will effectively end the release to the atmosphere.

Recompute the doses based on these assumptions. You will need to make changes on the release pathway screen to end the release, enter more forecast weather data, and increase the duration of the calculation.

Modify the latest meteorological data set to add to the Arkansas data the two additional forecast records shown below:

Type	Date	Time	Direction	Speed	Stability	Precip	Air temp
Fcst	today	23:00	230	2 mph	E	None	63 F
Fcst	next day	02:00	270	4 mph	E	None	59 F

Save the meteorological data set under a new name using the time: **1230**.

Run the calculations again and then fill in the table below:

Question	Your answer
What are the TEDE's at 2 and 5 miles?	
What are the thyroid doses at 2 and 5 miles?	
To what distance does the TEDE exceed the PAG?	
To what distance does the thyroid dose exceed the PAG?	

Inputs

Event Type   Nuclear Power Plant

Event Location   Arkansas - Unit 1

Source Term   Time Core Is Uncovered

Reactor shutdown: **10:00**

Core uncovered: **Yes, at 13:00**

Core recovered: **No**

Release Path   **Containment Leakage / Failure**

Release point characterization: **Not an isolated stack**

Release height: **0.0 m**

Consider building wake effects: **Yes**

Start of release to atmosphere: **13:00**

Release path events:

\<date\>	13:00	Sprays	Off
\<date\>	13:00	Leak rate	Design
\<date+1\>	01:00	Leak rate	0% per day

Meteorology   Data set type: **Actual**
                    Data set: **ARKA \<date\> 1230**

Calculation   Distance of calculation: **Close-in + out to 10 miles**.
Options
                    End calculations at: **Start of release to atmosphere plus: 15 hours**.

                    Rationale: This is the 12 hour release duration plus 3 hours for plume travel time to 10 miles.

Results

Question	Your answer
What are the TEDE's at 2 and 5 miles?	1.4 and 1.1 rem
What are the thyroid doses at 2 and 5 miles?	16 and 13 rem
To what distance does the TEDE exceed the PAG?	A little farther than 5 miles
To what distance does the thyroid dose exceed the PAG?	Between 7 and 10 miles

For very long release times

You may wonder how to handle a situation in which the release time is very long. For example, consider a reactor containment leaking at the design basis leak rate, which is likely to be under 1%/day. The leak rate is so small that the leakage will effectively be nearly constant over many weeks. Thus, you might think that you should end the calculations many weeks or even months after the start of the release. However, this is not the case. First, there is a limit of 48 hours on the calculation duration.

In addition, two other factors should be considered. First, the pressure in the containment will decrease with time as heat is removed. This will cause the release rate to decrease to lower than the rate used in RASCAL. Second, the projected doses being calculated by RASCAL are being used for early decisions on protective actions. When a release continues beyond a day it is no longer in the early phase of the emergency. RASCAL is designed primarily for early decisions that can be implemented before a release starts. Decisions about protective actions for long term releases can be made later as the situation becomes better understood and field measurements become available.

In conclusion, for the projected doses to be properly applied, the end of calculations for RASCAL should seldom exceed 24 hours from the start of the release. In fact, there is a limit on the calculation duration of 48 hours after the start of the release to the atmosphere.

SAVE THIS CASE!!
For consistency in the workbook, use the name: **Using RASCAL - Case 4**

12 Displaying Plume Footprints

Purpose

To learn how to obtain a picture so that plume direction and area exceeding PAGs can be evaluated.

Problem

Now we would like to see the spatial extent of the doses and the direction of the maximum doses.

Results

To do this we need to generate plume plots. Begin by selecting the **Detailed Results** button on the left side of the screen.

This will display the **Detailed Results of Dose Calculations** screen. The screen is divided into four main sections: Display Format, Result Type, Time Period, and Display Units.

Look first at the **Display Format** section:

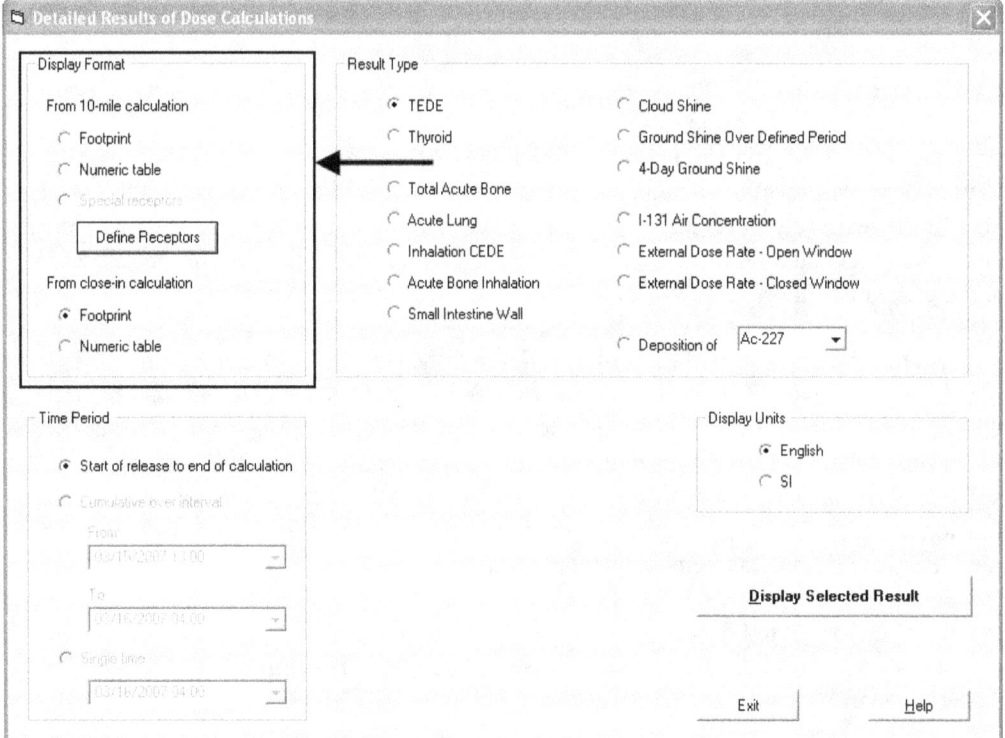

This section controls how the information is presented. Results can be displayed as either a numeric table or as a graphic "footprint". We must also choose whether we want to see the results from the close-in calculation or from the 10-mile calculation. We must look at them separately.

The **Result Type** section contains an option button for each of the 14 results calculated by RASCAL. Note that we have additional result types not shown in the maximum dose values table.

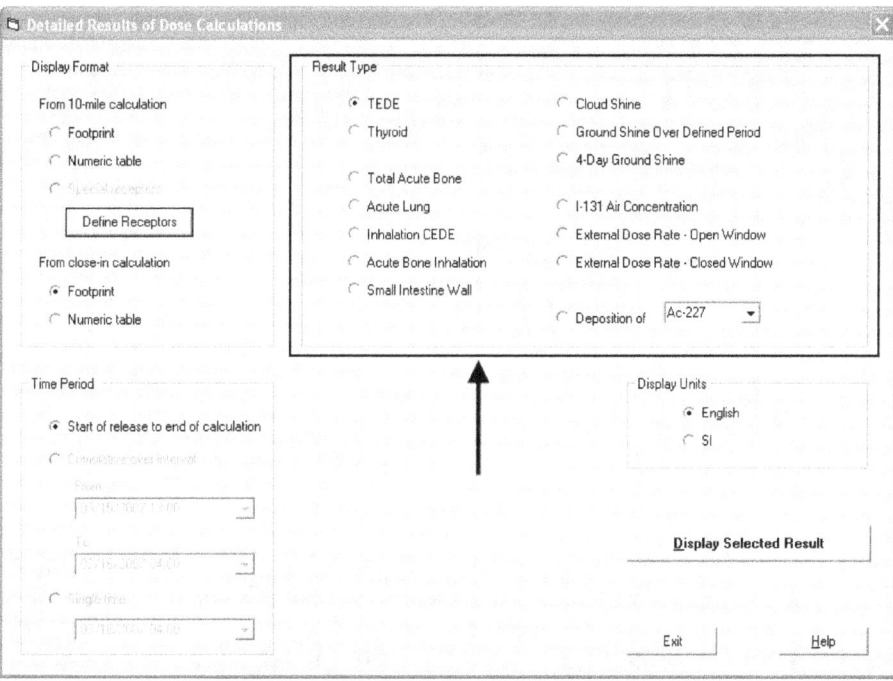

To get a footprint of the thyroid CDE from the close-in calculation, select the following options:

- Display format: **From close-in calculation - Footprint**
- Result type: **Thyroid**
- Time period: **Start of release to end of calculation**
- Display units: **English**

Select the **Display Selected Result** button.

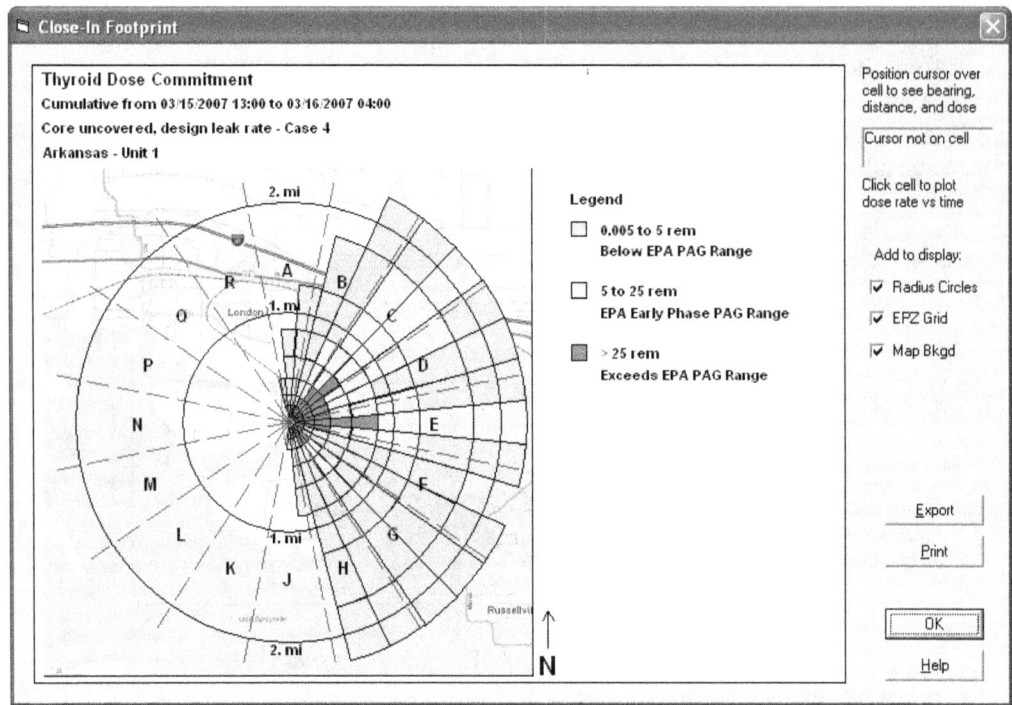

Note that the close-in footprint shape reflects the type of receptor grid used by the straight-line Gaussian plume model. Each colored area represents the dose at a model receptor point on the polar grid.

Click **OK** to return to the **Detailed Results** screen. Now display the thyroid dose footprint for the 10-mile calculation. The footprint for the 10-mile distance reflects the rectangular grid used by the puff model.

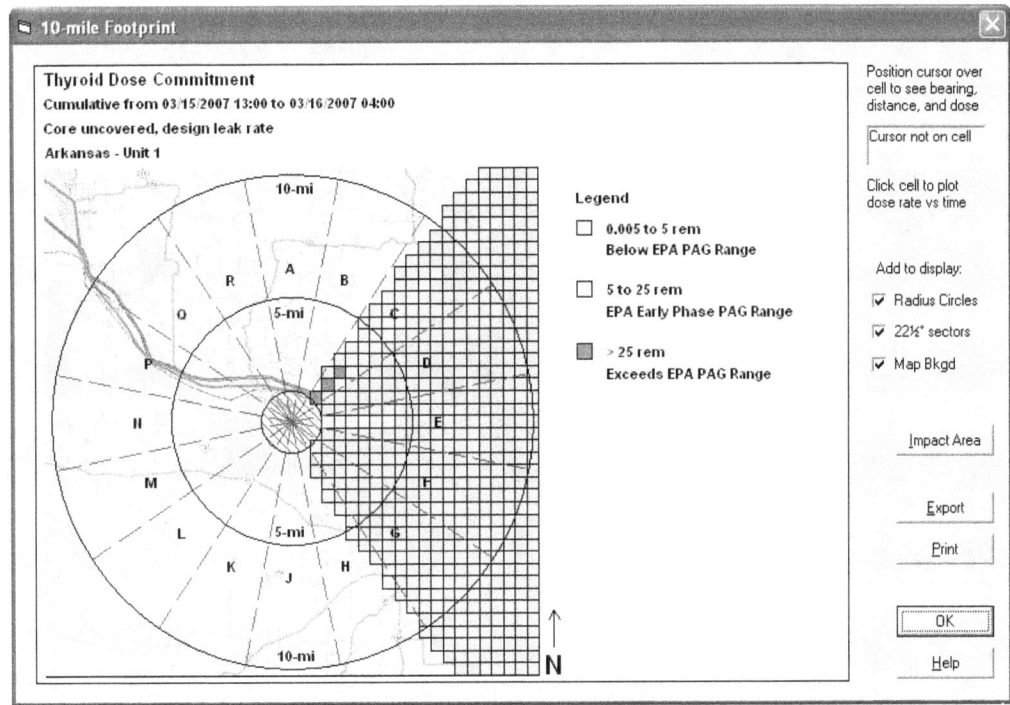

Answers the question:

Question	Your answer
In which compass direction from the release point are the doses greatest?	

Now, let's use the cursor to get some more detail about one of the individual cells on the footprint. You can see the location information and exact dose at any point and also examine the time history of the dose at a point.

Put the cursor in a colored cell of either footprint and note that the dose and location are displayed at the upper right of the screen. Follow the "centerline" out from the release point until you see:

Position cursor over cell to see bearing, distance, and dose

45.0°, 7.1 mi
7.56E+00

Here is the thyroid CDE of 7.56 rem at 7.1 miles northeast (45°) of the release point.

Click cell to plot dose rate vs time

Click the mouse while the cursor is positioned over that cell to display a rate plot.

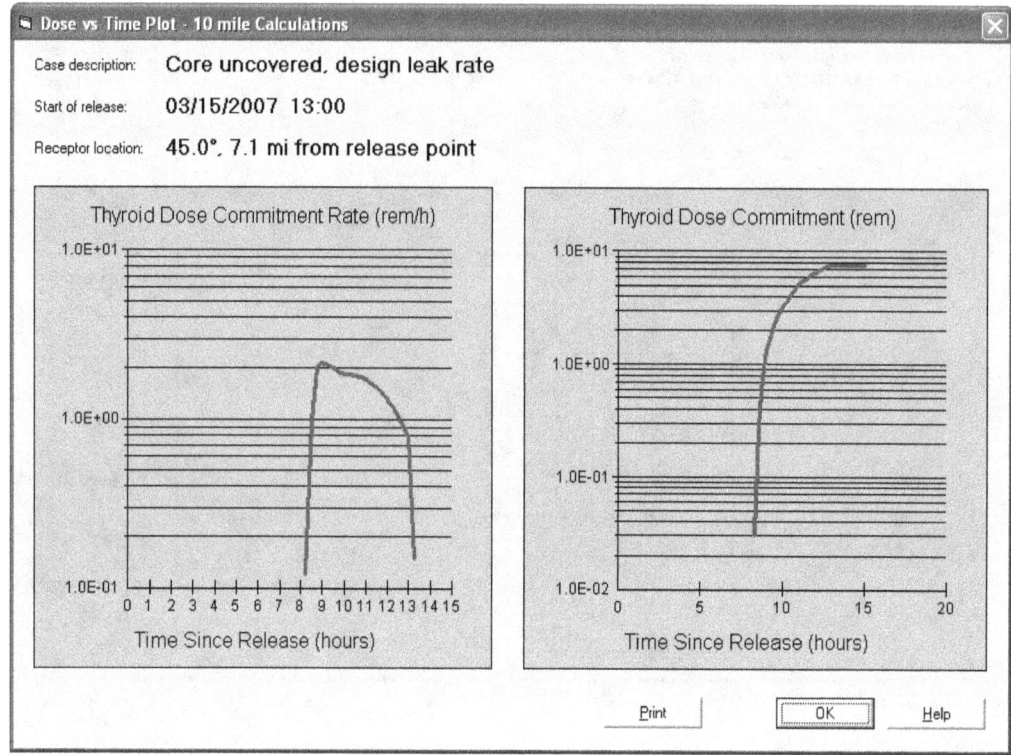

These graphs show us that 7.1 miles from the release point: the thyroid dose does not start to be delivered until about 8 hours after the start of the release, and continues to deliver the dose until about 12 hours after the start of release.

Did you get the answers below?

Question	Your answer
In which compass direction from the release point are the doses greatest?	To the northeast

13 Comparing Field Measurements with RASCAL Results

Purpose

To learn how to determine if the RASCAL results are consistent with a field measurement.

Discussion

Detailed results allow not only varying the result type but examining doses and dose rates at intervals during the modeling duration. For example, instead of looking at thyroid doses for the entire 15 hour period of the previous case, you could look at the thyroid dose over an interval in the middle of that 15 hour block.

RASCAL calculates external dose rates for both open and closed window meters. These are useful for comparison with field readings.

Problem

Assume that the accident has unfolded as expected. We are in the middle of the release and a field team reports a closed-window (gamma) meter reading of 1.0 mR/h at 5:30 P.M. The measurement was made on the approximate plume centerline at a distance of 5.5 miles east of the release point.

How does this measured value compare to the value predicted by the RASCAL source term to dose model?

Results

To get the reading we need for comparison with the field data, make the following 2 selections on the detailed results screen.

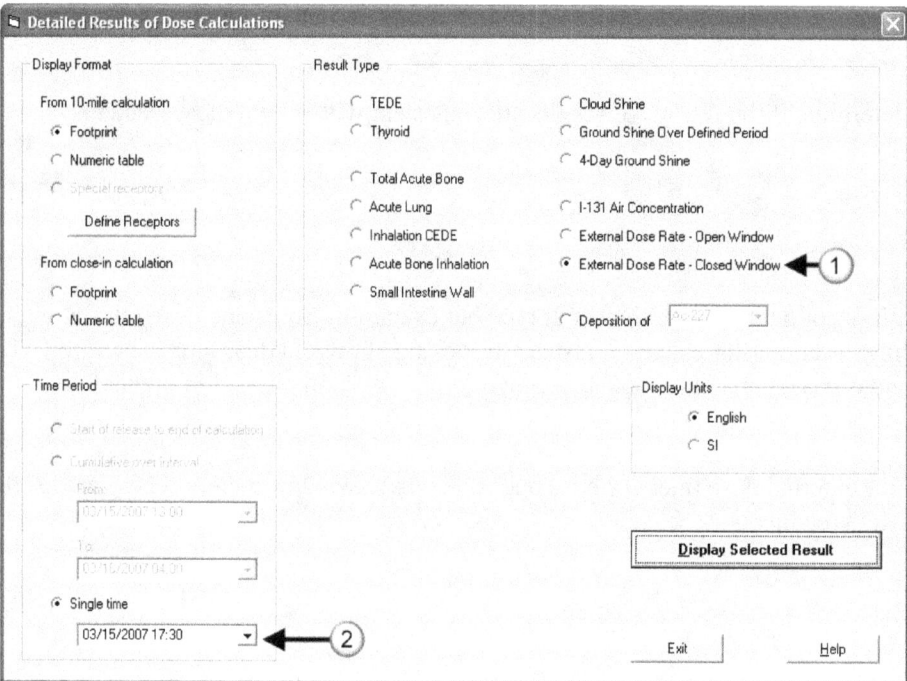

1. For result type, select **External Dose Rate - Closed Window**. This will display the gamma dose rate.

2. For time period, only the single time option is available. Select the time to match the field reading (**17:30**).

Click the **Display Selected Result** button.

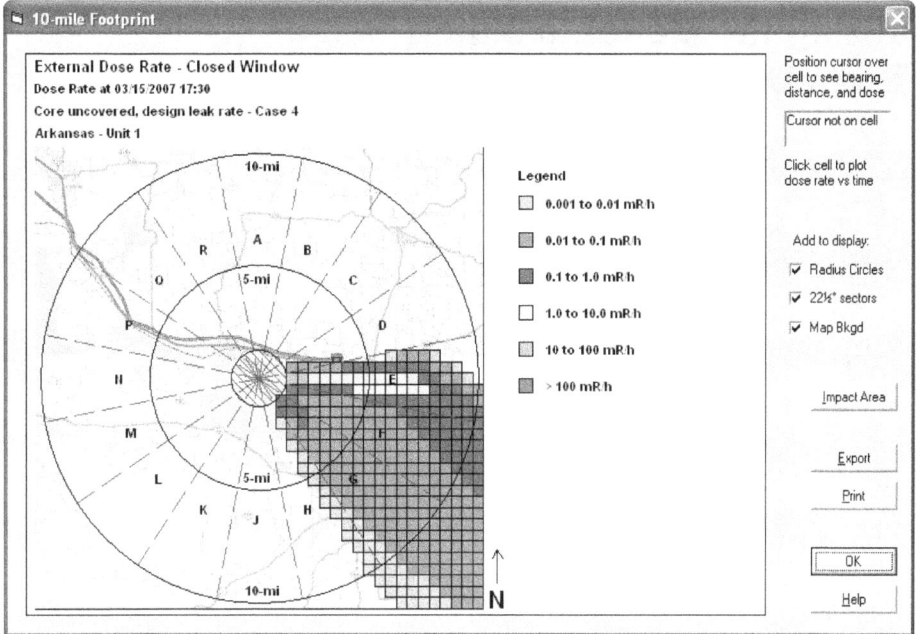

We are interested in the gamma rate at a point 5.5 miles east of the release. Move the cursor over the colored cells and watch the display in the upper right. As you traverse the yellow cells at some point the box will show 90° at 5.5 miles.

Position cursor over
cell to see bearing,
distance, and dose

90.0°, 5.5 mi
4.00E+00

The 4.00 is the gamma dose in mR/h at that location.

Conclusion

These results are pretty close considering all the uncertainties involved. Be sure to confirm with multiple readings. However, we should feel comfortable about the modeling run. Tell supervisors and decision makers that the field readings are consistent and add confidence to the model results.

14 Normalizing the Model to Match Field Team Data

Purpose

To learn how to normalize RASCAL Source Term to Dose model results to match field team data.

Discussion

The basic method for normalizing the model results consists of two parts: obtaining the field measurements and then adjusting the model runs.

Obtain field measurements

1. Use 1 or 2 field teams to traverse the plume at least once and preferably twice. It is necessary to actually *traverse* the plume. Field measurements that do not include measurements near the plume centerline are essentially useless. If all the measurements are near the plume edge, you will learn if there is something there or not but you cannot estimate the magnitude of what is there. You cannot determine if your RASCAL results are too large, too small, or about right.

2. The easiest type of measurement to get is a gamma (closed-window) exposure rate reading in mR/hr. The gamma rays will come from two sources: radionuclides in the plume and radionuclides that have been deposited on the ground. Early in the release most of the gamma rays will come from airborne radionuclides in the plume. Late in the release, gamma rays from radionuclides deposited on the ground may predominate. Your meter will not be able to distinguish the two gamma ray sources.

3. Measured gamma dose rates are expected to vary greatly over brief time intervals and over short distances. Our models represent airborne concentrations as slowly and steadily decreasing with distance from the plume centerline. The model is a good representation of reality only if measured concentrations are averaged over some time, for example 5 to 10 minutes. To avoid being misled by very short-term fluctuations, you must do one of two things. You must note the dose rate over a period of several minutes and record the average dose rate during that time interval, or you must make several (at least 4 or 5) nearby measurements and average them. Making several nearby measurements separates your measurements in both time and space.

4. Assuming that your gamma measurements have been taken so that the appropriate averaging has been done, you will still need measurements for several locations near the plume centerline. In addition, you should take measurements to roughly estimate the location of the edge of the plume.

5. At each measurement point, record:

   a. - the closed-window meter reading in mR/h
   b. - the time the measurement was taken
   c. - the location of the measurement point so that you can locate it on a map.

6. Another type of measurement that you can get is an air sample. Air samples are normally taken over some time interval such as 5 minutes so they do average the concentration over a time interval.

However, air samples are harder to get than gamma rate measurements and they take longer to analyze.

RASCAL Modeling Runs

1. Make a RASCAL run using the best estimate of source term, release, and meteorology. Your run should be a best estimate. You will learn very little if you do a worst case run and find that the worst case RASCAL case gives larger doses than the field measurements. Run the calculations until plume passage would be complete.

2. Select the **Detailed Results** button. You will get the screen shown below.

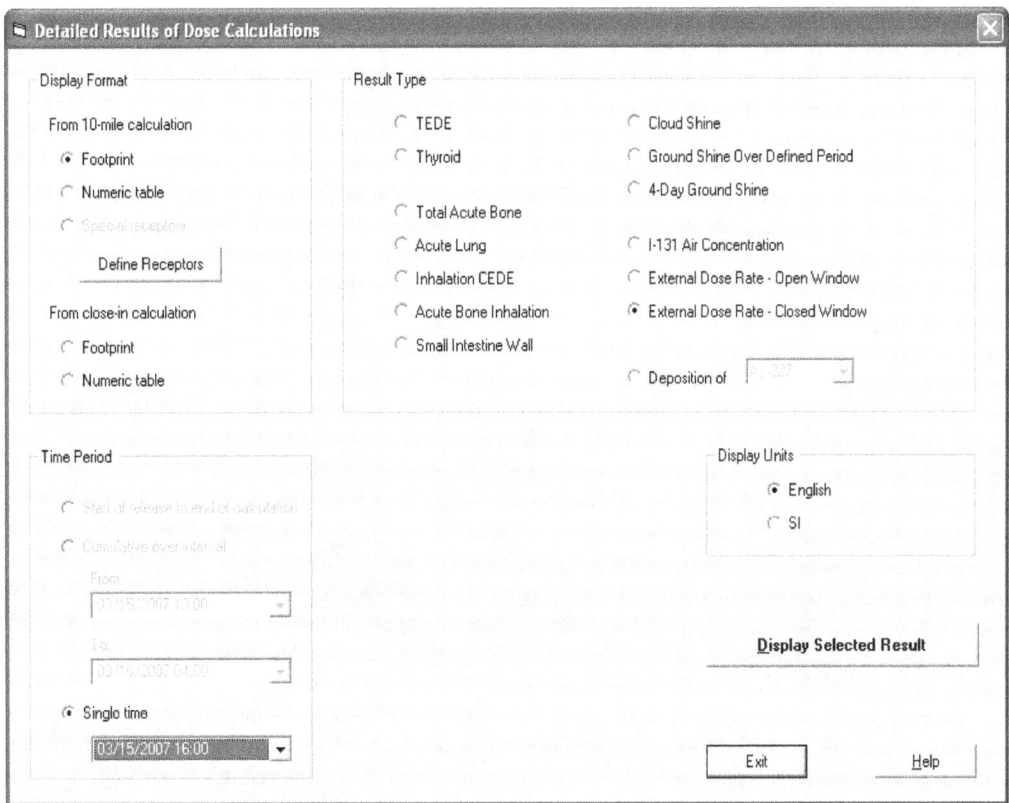

3. For Result Type, select **External Dose Rate - Closed Window**. For the Time Period, only the "Single time" option will be available to you (because dose rate must be associated with a single time). Select the time that most closely matches the field measurement time. (If your field measurements were taken over widely varying times, you will have to look at several sets of RASCAL results, each corresponding to the appropriate field measurement time.)

4. Under Display Format, determine if your field measurement location is "close-in," usually less than 2 miles, or further out. You have the choice if you want to see the External dose rate - closed window results as a "Footprint" (picture) or as a numeric table. Either will work, but we recommend the Footprint because it shows the map background, which will make your job easier.
5. With the footprint displayed, place your cursor on the displayed plume at the distance and direction from the release point corresponding to your field measurement. The exposure dose rate, the distance,

and the direction are displayed in the box at the upper right. The dose units are in the legend. Record the RASCAL exposure dose rate. Repeat for each measurement.

6. Compare the RASCAL footprint results with the field measurements. Which is larger and by how much? Is the location of the plume centerline from the footprint and the field measurements about the same? Are the edges of the plume in about the same place?

Problem

Assume that the accident is unfolding as expected. Make sure that **Case 4** calculations are loaded.

Field teams have traversed the plume. The roads used were 5-6 miles away from the release point and were to the east of the release.

The following measurements of gamma exposure rate were reported.

ID	Time	Direction from release point	Distance	Measured	RASCAL result
		degrees	mi	mR/h	mR/h
S1	17:18	79	6.0	0.05	
S2	17:22	86	5.5	0.7	
S3	17:25	91	5.7	8.2	
S4	17:30	94	5.1	4.9	
S5	17:34	99	5.9	1.9	
S6	17:41	105	6.1	0.12	

Recording RASCAL Results

We are going to first direct our attention to the centerline dose, which is measurement S3 with a reading of 8.2 mR/h. Seeing a reading of 4.9 mR/h for measurement S4, which is about the same magnitude gives us confidence that our highest measured dose is not an outlier.

Use the **Detailed results** screen to select the dose type, distance, and time period. The dose type should be the **External Dose Rate - Closed Window** to see the needed gamma exposure rate projections. The **Footprint** format will provide an easy to use method for picking off doses. Since we are interested in doses at a distance of 5-6 miles we need to use the **From 10-mile calculation** display format. Set the time period to **Single time 17:30** to correspond to the general time when the measurements were taken.

Click **Display Selected Results**. Now use the cursor (as described in the previous problem) to obtain the doses from cells on the footprint display that are closest to the measurement locations. Fill in the table above with the model results.

Results

The following RASCAL results could be picked off the footprint display.

ID	Time	Direction from release point	Distance	Measured	RASCAL result	
		degrees	mi	mR/h	mR/h	(deg, distance)
S1	17:18	79	6.0	0.05	0.0027	80.5, 6.1
S2	17:22	86	5.5	0.7	0.81	84.8, 5.5
S3	17:25	91	5.7	8.2	4.0	90, 5.5
S4	17:30	94	5.1	4.9	2.1	95, 5.5
S5	17:34	99	5.9	1.9	0.25	100, 6.1
S6	17:41	105	6.1	0.12	0.05	104, 6.2

Interpretation of the Comparison

These are our recommendations for interpreting the comparison:

1. Don't put too much weight on just a few field measurements taken within a short time interval and from a few locations in close proximity. If one set of field measurements is not matching the model, get additional sets of measurements before you draw strong conclusions.

2. If the centerline doses are within a factor of 2 or 3, that is dead on. You can say that the field measurements agree with the model results.

3. We would start to get a little concerned if the field measurements were more than 3 times higher than the model predictions. RASCAL is designed to give a somewhat conservative (over-estimate) of doses. If field measurements were 3 times higher than predicted, we would start to worry that perhaps we are not understanding what is happening and perhaps recommended protective actions are not adequate.

4. If the field measurements are a factor of 3 to 10 lower than the model results, we would probably not adjust the model inputs to rerun the model nor would we change a prior protective action recommendation. We would probably tell the decisionmaker that the field measurements were a little lower than the model predictions, but are roughly consistent. We would definitely not quantify the size of the difference to the decisionmaker because we don't want to get caught up in details and trivia.

5. If the field measurements are coming in more than a factor of 10 lower than the model predictions, this indicates that our model inputs need to be adjusted. (This assumes that we have confidence in the field measurements.) We would ask, where could we have gone wrong? We would probably adjust the model inputs and rerun the RASCAL model.

6. If the direction of the maximum doses (presumably the plume centerline) don't agree (meaning more than 10 degrees), it means the model wind direction may be off from the actual wind direction. We would not try to adjust the meteorology, but we would add a qualifier to the results describing the difference.

15 Field Measurement to Dose

Purpose

To learn how to use the Field Measurement to Dose model to calculate intermediate phase doses based on measured ground concentrations.

Discussion

The Field Measurement to Dose model would normally be used after releases had ended to determine whether the area is habitable or do residents need long-term relocation. The model estimates doses based upon measurements of ground contamination. Within the NRC, the site team is the most likely user. The calculations can also be done using the Turbo-FRMAC code.

Two EPA pathways are considered: groundshine and inhalation due to resuspension. Both pathways are affected by weathering. The material becomes less accessible through time as natural processes work. The EPA sets the following limits:

1st year PAG = 2 rem

2nd year objective = 0.5 rem

50 year objective = 5 rem

Problem

The release from Arkansas Nuclear One has stopped. Field teams have been dispatched to determine where and how much radioactive material was deposited. The first report to come in provides:

Sample ID: ABC123

Location: a street corner in a subdivision downwind of the plant

Sample analyzed: day after accident at 13:55

Sample data

Nuclide	Concentration microcuries/meter^2
Ba-140	1.4
Cs-134	0.6
Cs-137	0.35
I-131	1.7
I-132	1.3
La-140	1.3
Sr-89	1.0
Te-132	1.2

The residents were evacuated early in the accident. Is it now safe for them to return to their homes?

Inputs: Step-by-step

Start the **Field Measurement to Dose** model from the main RASCAL screen.

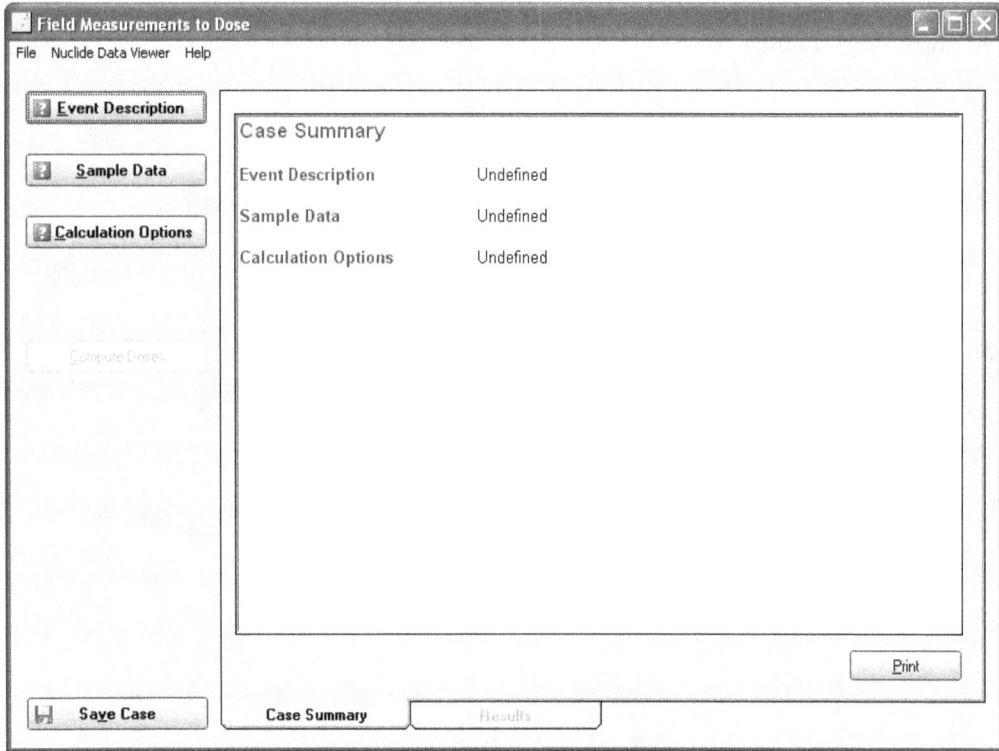

Operation of the user interface is similar to the Source Term to Dose model; buttons on the left, top to bottom. A tab area shows the summary of inputs and the results of the calculations

Begin by selecting the **Event Description** button on the main screen. That will display the following.

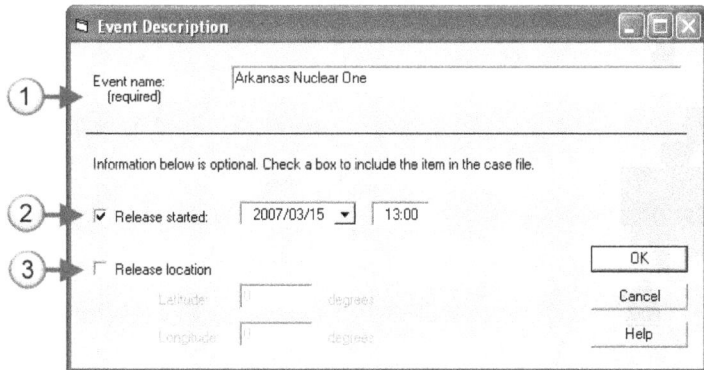

1. Event name: **Arkansas Nuclear One**

    An entry is required but the text is used for labeling only; it has no impact on the calculations.

2. Release started: put a check in the box, change the date if needed, and enter a time of **13:00**

    This information is optional and is used only for labeling purposes.

3. We could also enter the latitude and longitude of the release point. For this case it would be the information reported in the Source Term to Dose model case summary for Arkansas.

Click **OK** to return to the main screen. Note that the **Event Description** button now has a checkmark and the case summary has been expanded.

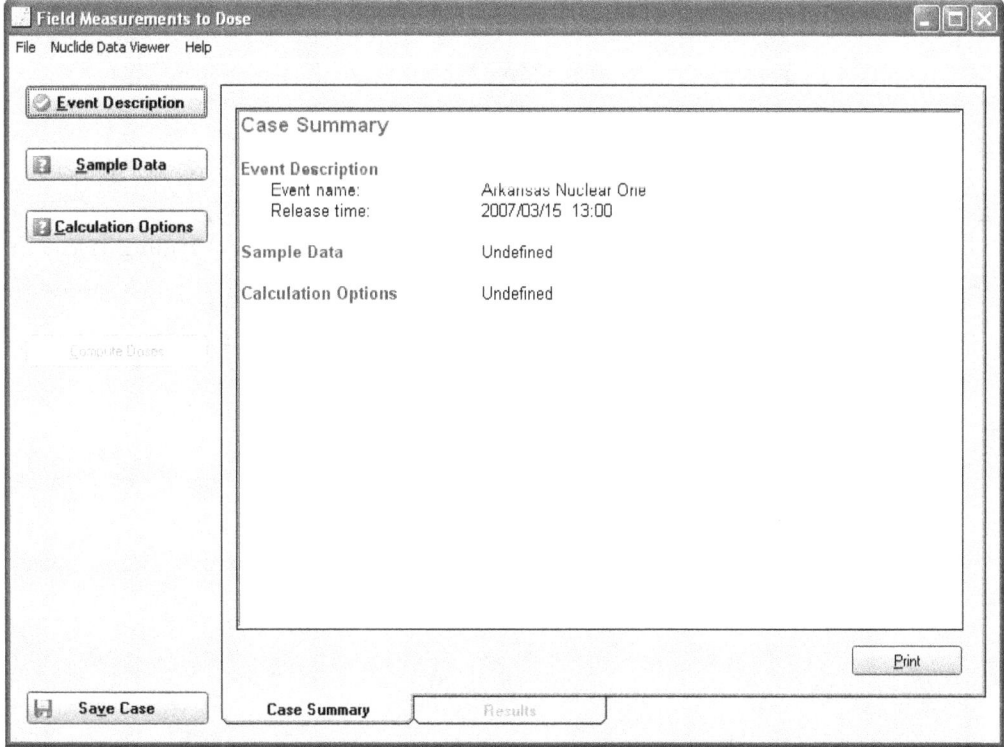

Next, select the **Sample Data** button.

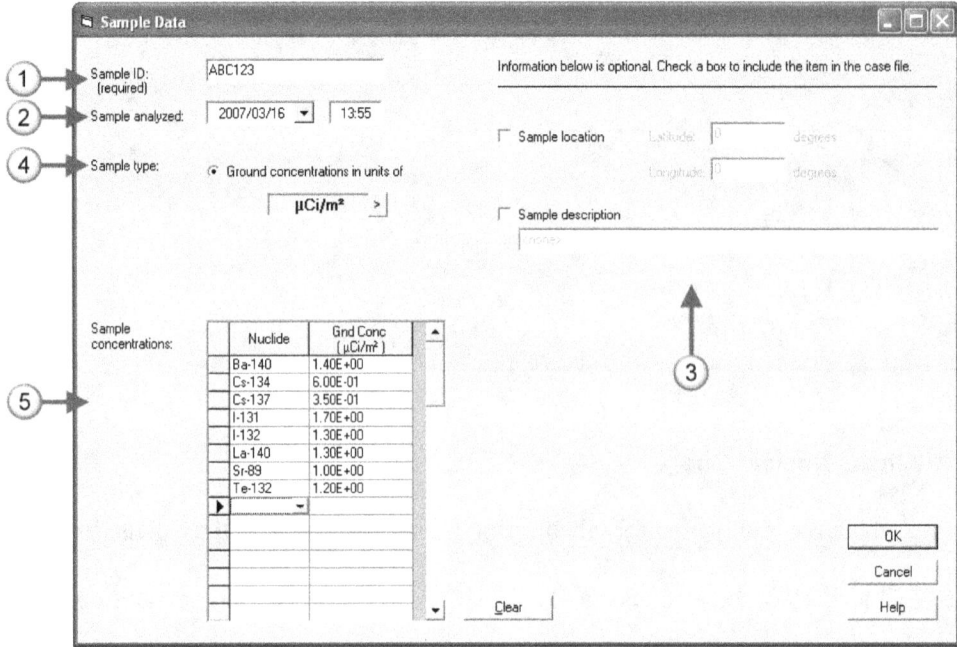

1. Type in the sample ID provided: **ABC123**

   This entry is required and should uniquely identify the sample. It will be used to label the results.

2. Set the date and time the sample was analyzed: **<date+1> 13:55**

3. If provided, you may optionally describe the location of the sample and/or the type of sample.

4. The default ground concentration units of **μCi/m²** are correct so no change is needed.

5. Enter the nuclide names and concentrations for the sample.

   You can speed up the data entry process by typing in the nuclide name without using the drop-down list and then tabbing to the next column to enter the concentration. Tabbing again gets you to the next row ready for the next nuclide name.

When the sample data is complete, click the **OK** button.

Finally, from the main screen, click the **Calculation Options** button.

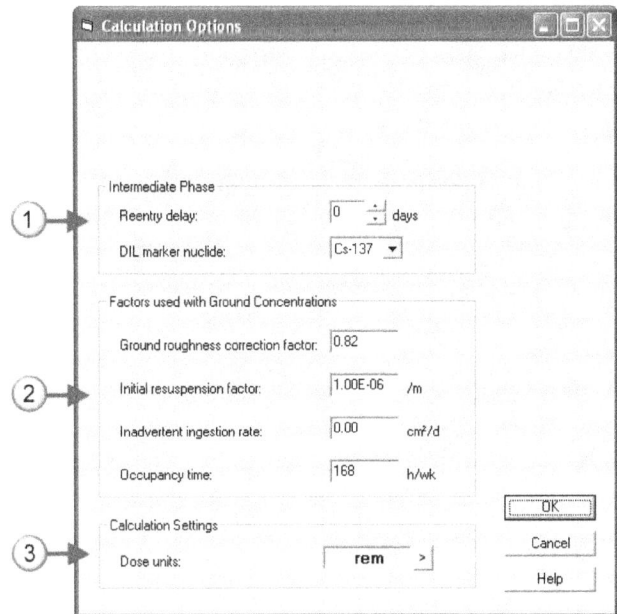

1. Leave the reentry delay at **0** days. Select **Cs-137** as the DRL marker nuclide.

    The reentry delay is used for areas in which people have evacuated. For example, consider a case in which the measurements and analysis needed to define a relocation area will take a week. The EPA PAGs are based on the concept of averted dose - the dose that can be avoided by taking an action. To determine if an area is suitable for reentry a week after the accident, the dose that could be averted by not reinhabiting the area is compared to the EPA IP protective action guides. Thus, the first year PAG would be compared to the dose not including the first week.

    In some instances it may be difficult to use the exposure rate DRL to identify areas where doses might exceed the protective action guides. Examples are when the exposure rate is near background levels or when there are no gamma-emitting radionuclides in the mix. In those instances, it may be easier to measure the surface concentration of a particular marker radionuclide rather than the exposure rate. You may select the marker nuclide to be used from the list of measured nuclides on the ground.

2. Leave the four factors at their defaults.

    The ground roughness correction factor accounts for the fact that the ground surface is not perfectly smooth. Some of the ground shine radiation will be intercepted by this uneven surface and cannot contribute to the dose. EPA used to use a value of 0.7 but now we use 0.82.

    Intermediate-phase inhalation doses are calculated assuming activity on the ground is resuspended and then inhaled. The resuspension process in the model begins with the user entered resuspension factor and decreases with time. The default initial resuspension factor is **1.00E-06** /m and represents normal conditions for a temperate climate area. The value should be changed only if there is reasonable confidence that resuspension rates will be different. For example, with a very dry and windy area, the factor might be increased.

    Inadvertent ingestion is the accidental ingestion of radioactive material from contamination on surfaces. This can happen for example if a person touches a contaminated surface picking up a small amount of radioactive material on their hands. If they then hold a sandwich, some of the material will transfer to the sandwich and be eaten with the sandwich. The default inadvertent ingestion rate is **0.0**, assuming no contribution to the intermediate-phase dose. Neither EPA or FRMAC will consider inadvertent ingestion in their calculations.

The occupancy time describes how long a person would be exposed. The default of 168 hours per week assumes the worst case - they are there the entire time.

3. Set the dose units to whatever is most useful. SI units can be used if desired.

Click the **OK** button to return to the main screen again.

Now the problem is fully defined and the program is ready to calculate the intermediate-phase doses.

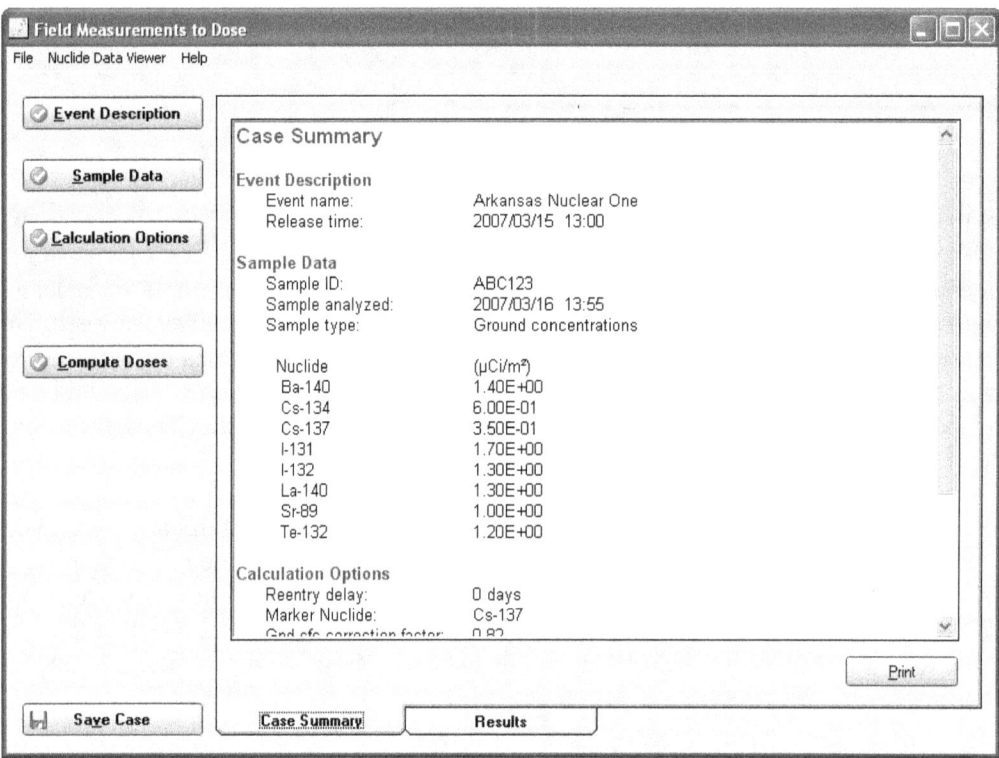

Click the **Compute Doses** button. The results will automatically be displayed.

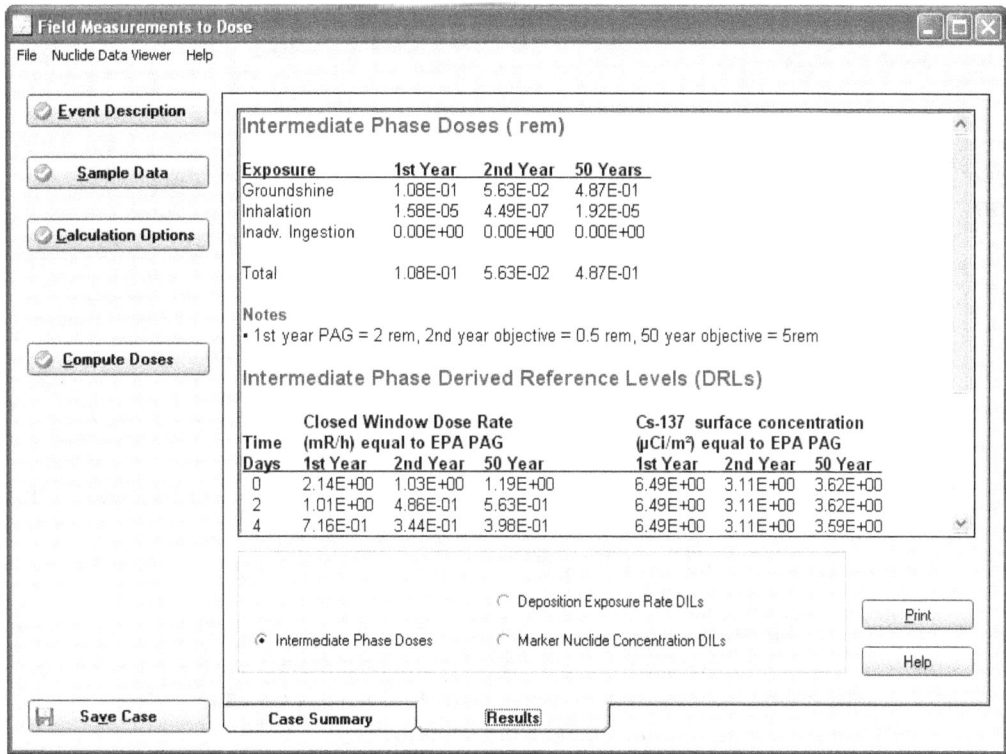

Results are far below the EPA PAGs for the intermediate phase. Based on that single result, people can immediately return to their homes in the area. However, measurements at only one location and/or only one sample is *not* enough data to make a decision.

For a typical "reactor mix" of radionuclides, if the first year dose is below the PAGs then the second and fifty years doses will be below PAGs as well. Also, the groundshine dose will usually dominate.

16 Briefing for Non-Technical Decision Maker

Purpose

To learn how to prepare a briefing for a non-technical decision maker.

Discussion

Whether you will ever brief a non-technical decision maker or not, you need to understand the information that should be in the briefing. Until you know which information is important to convey and which information is not useful, you will not be able to do a good job of explaining your RASCAL results.

In the briefing of a non-technical decision maker, the numbers and the technical terms that we have been using in this course are not appropriate for the briefing. That information is too technical and detailed. The decisionmaker needs the big picture. The public needs to be told what it needs to do to escape from danger. They do not want and will not understand technical details.

We are going to give you insight into what information should be conveyed and what information will just confuse the message.

Problem

Arkansas Nuclear One, Unit 1, declared a General Emergency at 10:20 A.M. Assume that the standard protective action recommendation for a General Emergency is evacuation to 2 miles full circle and to 5 miles downwind in a 45° sector.

Your job is to evaluate the appropriateness of this standard protective action recommendation and prepare a briefing for a non-technical decision maker that describes the adequacy and appropriateness of the standard protective action recommendation for the specific accident being evaluated.

Assume that the person making the public statement will be the Governor of Arkansas. Your briefing should prepare him to make a public statement. Therefore, your briefing should include the basis for the decision in language that is understandable to a non-technical person.

Assume that there are no site-specific protective action sectors or factors that must be considered.

Points to Include in Briefing the Decision Maker

The points below should probably be included in the briefing. The problem you have been working did not include detailed information of the approved emergency plan for the site or site-specific information so you cannot include that information in this exercise briefing. However, in a real event, that information should also be included.

Topic	Rationale
Existing emergency plan: There is an existing State emergency plan for handling accidents of this type. The plan was carefully thought out and reviewed and approved by many state departments and by federal agencies.	You want to reassure both the decision maker and the public that you know what you are doing. The fact that you have an existing plan that was carefully thought out is a big plus for credibility.
The plan has a solid basis: The emergency plan is based on careful consideration of the potential consequences of nuclear power plant accidents and on guidance established by the U. S. Environmental Protection Agency (Manual of Protective Action Guides and Protective Actions for Nuclear Incidents, USEPA, 1992).	The public will be reassured if they believe that the situation was carefully analyzed beforehand. The EPA is an agency with credibility for protecting the health and safety of the public.
Balancing benefits against detriments: The radiation dose levels for taking protective actions that are established in the EPA guidance were developed by balancing the potential benefits achieved by taking the actions against the potential detriments of taking the actions.	The decision maker should realize that the PAGs balance benefits against detriments. Any political decision maker will be aware of the detriments to a large scale evacuation. The decision maker needs to be assured that those detriments were considered in establishing the PAGs.
You are following the plan: Recommend that the decision maker follow the existing plan and order an evacuation as specified in the plan. In this problem that calls for evacuation to 2 miles full circle and 5 miles downwind.	Both the decision maker and the public will be reassured that the response actions were carefully thought out and planned in advance.
Maps of evacuation areas: Have a map of the areas showing the areas that you think should be evacuated. Do NOT show a plume plot to the decision maker.	You need a map. Trying to describe an evacuation area without a map would be incomprehensible to anyone.
Maps showing plume plots: NEVER, NEVER, EVER show a plume plot to a decision maker.	Plume plots are too technical and confusing for a non-technical decision maker. They are for you, not the decision maker. You already translated the plume plot into a recommended evacuation area. That's all you have to tell the decision maker.

Topic	Rationale
Description of evacuation area: Say something like: the downwind direction is _____. The width of the downwind evacuation should be (select whichever is appropriate) (1) 45 degrees because the wind direction is well-known and persistent, or (2) ____ degrees because the wind direction is expected to change during the release or the wind direction is uncertain.	Use the results that you have from the problem.
Put the calculated dose plots in their proper context: Say something like: our calculations of projected doses suggest that the above evacuation is significantly larger than would be needed based on the EPA guidance. Thus, a smaller evacuation could be justified. However, we believe that, as a general principle, existing plans should be followed. According to our dose projections, the evacuation according to the plan is more than adequate to protect public health and safety.	The decision maker should understand the basis for your recommendation, but he does not need all the technical details. Avoid technical jargon. Talking about TEDE, rems, curies/square meter, projected doses, and 4-day groundshine will just confuse him.
Real measurements: We have sent radiation monitoring teams into the field to measure radiation levels. At this time no release of radiation has been detected. We will continue to have monitoring teams in the field to measure radiation levels.	Real measurement data beat theory any day. The fact that you are really checking what is really happening is reassuring. Even zeros as results are great.
Describe the current situation: For this problem you might say something like: there has been no significant radiological release at this time. The evacuation being ordered is a precautionary measure to guard against the possibility of a significant radiological release.	Be very careful about how you describe the release and the reason that the evacuation is recommended.
Urgency: For this problem, you might say something like: the evacuation should be prompt but orderly. Current conditions are not dangerous.	The public needs to be told the truth about the degree of urgency needed for their actions.
Return to homes: Inform the decision maker that there is a possibility that residual radioactive contamination in some areas may (or may not) make those areas uninhabitable in the future, but that cannot be determined until the incident is over and measurements of deposited radioactivity have been made. The decision maker should say that a decision on return will be made when we are sure it is safe to return. We need to determine that the plant is stabilized and that radiation levels, if any, are below established guidelines.	People need to be informed of the possibility of long term relocation. If they become aware of that only later, they will feel they were not told the whole truth.
Children in school: Describe how the state plan deals with the issue of children in school.	This will be of great concern to parents. It must be described in public announcements.

Topic	Rationale
Use of potassium iodine: If you expect that people will evacuate before the plume reaches them say something like: the use of potassium iodine for the public is not recommended at this time because the evacuation should keep thyroid doses well below the levels for which the FDA recommends use of potassium iodine.	The FDA has established criteria for the use of KI. The state plan will have considered the FDA recommendations. Follow the state plan.
Pets: Inform the decision maker on how the plan deals with the issue of people's pets.	People will be concerned about their pets. They must be told what they should do. They will be reassured if they are aware that it has been considered.

Nuclear Power Plants

The section "Using RASCAL: Assessing a PWR Core Damage Accident" ran through the entire code for one particular reactor core damage accident to teach you many of the details of the code. This section on Nuclear Power Plant Problems has short problems, each illustrating some specific aspect of the code. These problems primarily explore reactor source terms and release pathways.

17 Examining the RASCAL Source Term

Purpose

To learn how to look at the details of the source term generated by RASCAL and how to export the source term information to the IMAAC.

Discussion

Section 3 described how RASCAL calculates the source term used in the analysis.

The October 2004 Nuclear/Radiological Incident Annex of the National Response designates the Interagency Modeling and Atmospheric Assessment Center (IMAAC) as the single source for atmospheric dispersion and consequence prediction. During a radiological Incident of National Significance in which the NRC is the Coordinating Agency, RASCAL may be used to supply a source term to the IMAAC.

Source terms should be sent to IMAAC *only from NRC Headquarters*, not from the NRC Regions or the NRC site teams.

Problem

Using the case saved at the end of Problem 6 (should be Case 1),

1. Answer the following questions about the source term:

    Of the total activity released to the atmosphere, what fraction is noble gases?

    Iodines are always a concern in a release. Describe the time dependence of the iodine release. At what time is the iodine release rate the greatest?

2. Prepare a source term export file for sending to the IMAAC.

Results

Load the saved case from problem 6, the suggested name was **Using RASCAL - Case 1**. You can do this by selecting **File | Open Case** from the menu bar and then selecting the file name from the list.

Select the **Source Term Summary** tab.

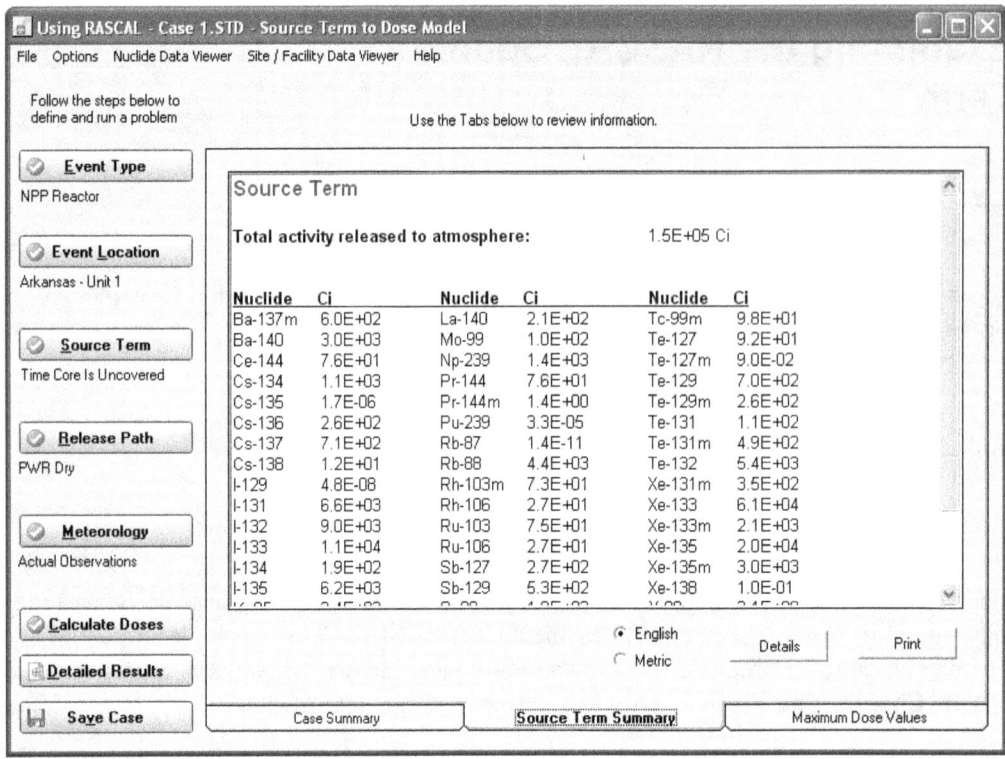

The total activity (1.5E+05) is given at the top of the summary. To get the noble gas activity fraction, you will have to manually add up the activities for all the noble gases and then divide by the total activity released to the atmosphere.

Noble gas	Activity (Ci)	
Kr-85	3.4E+02	
Kr-85m	3.3E+03	
Kr-87	7.4E+02	
Kr-88	5.6E+03	
Xe-131m	3.5E+02	
Xe-133	6.1E+04	
Xe-133m	2.1E+03	
Xe-135	2.0E+04	
Xe-135m	3.0E+03	
Xe-138	1.0E-01	
Total	**9.64E+04**	**9.64E+04 / 1.5E+05 = 64.3 %**

For reference, the radionuclide releases from TMI and Chernobyl are show below.

Nuclide group	TMI (Ci released)	Chernobyl (Ci released)
noble gases	2.5 million	170 million
radioiodines	15	48 million
others	negligible	72 million

TMI reference: Table II-1 of "Three Mile Island: A Report to the Commissioners and to the Public," Nuclear Regulatory Commission Special Inquiry Group, Mitchell Rogovin, Director, 1980.

Chernobyl reference: Table 1 of "Chernobyl - Assessment of Radiological and Health Impacts," Nuclear Energy Agency, Organisation for Economic Co-Operation and Development, 2002.

To see the time dependence, select the **Details** button on the tab.

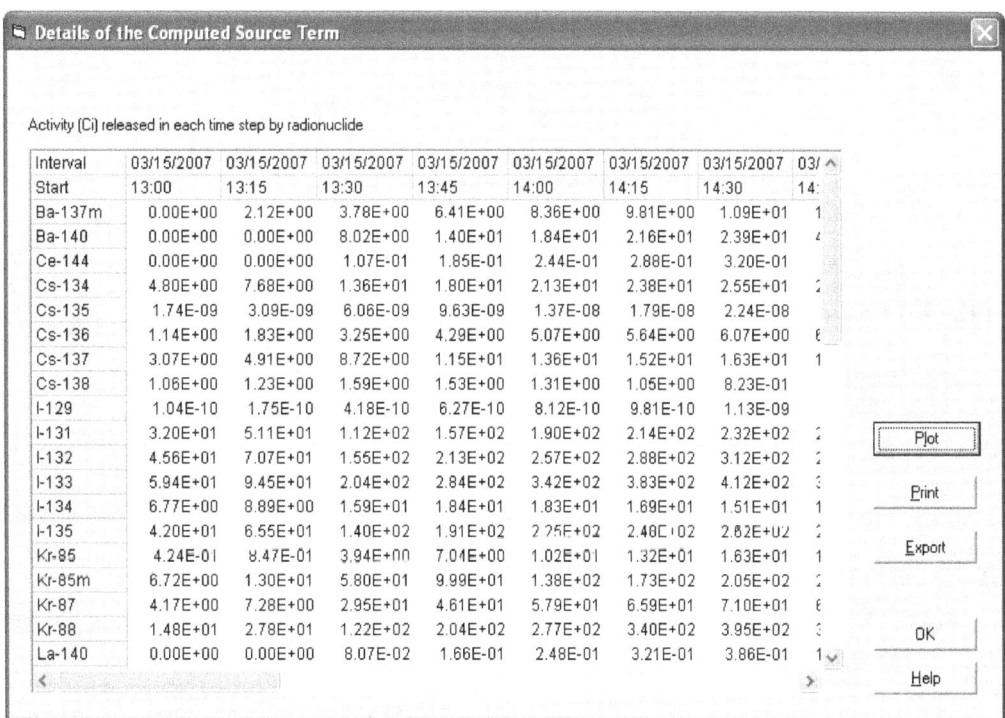

This shows the activity of each radionuclide released to the atmosphere each 15 minute time step. Examination of the iodine rows shows the amount released each time period increasing.

Next, click the **Plot** button. In the selection boxes on the right, select I-131, I -132, and I-133. Then, click the **Update Plot** button.

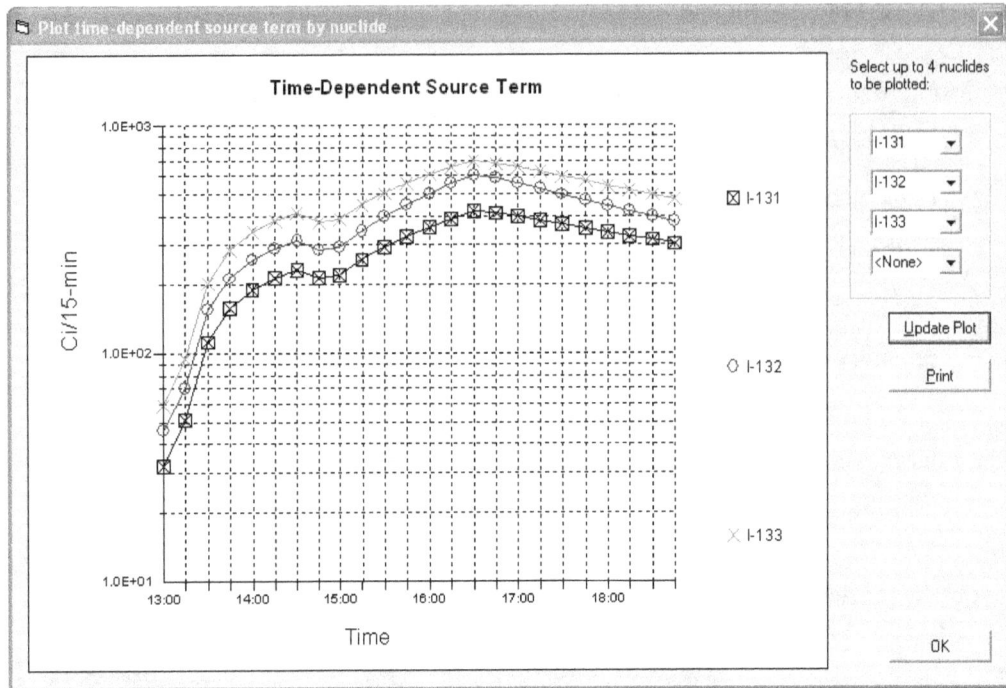

The graph shows the release rate of the selected iodines in units of curies per 15-minute time step. From the graph, the rate is greatest at about 16:30.

To create the IMAAC export file

1. Return to the "details" view of the source term.

2. Click the **Export** button to the right of the nuclides table.

3. The message explains that you are creating a comma-delimited text file. Click the **Yes** button to continue.

4. Navigate to the folder where you want to save the file. Enter a unique file name. Click the **Save** button.

5. The file can now be attached to an e-mail and sent to the IMAAC.

    NRC staff - Refer to Protective Measures Team procedures for IMAAC notification and additional information that needs to be in the e-mail. Source terms should be sent to IMAAC *only from Headquarters*, not from the NRC Regions or NRC site teams.

18 Release Pathway Reduction Mechanisms

Purpose

To study how to use release pathway reduction mechanisms to obtain an appropriate release.

Discussion

The magnitude of the leakage rate of radioactive material to the atmosphere is an important consideration when estimating doses. One way to define leakage is by a containment pressure and a containment hole size. Reducing either will reduce the rate at which material enters the atmosphere.

Various reduction mechanisms are modeled by RASCAL. The reduction of the source term by sprays is treated as an exponential function with time. There is an initial effectiveness that is applied for the first 15 minutes that radionuclides are in the containment atmosphere and then a reduced effectiveness acting after 15 minutes. The reduction factor (RDF) is calculated as:

$$RDF = e^{-\lambda T}$$

where T is the total amount of time (h) that the sprays have been operating and lambda (λ) is 12/h for the first 15 minutes and 0.2/h after 15 minutes.

Some PWRs use ice bed condensers to allow for smaller containment volumes. Such containments route hot gases through large ice baskets to condense steam and lower containment pressure. Factors to consider with this type of containment are whether the ice has been exhausted and whether the recirculation fans are available to provide multiple passes of the hot gases through the ice beds.

Problem

The Catawba Nuclear Plant, Unit 1 (a PWR with an ice condenser containment) had been operating at full power. The reactor scrammed at 10:45 A.M. local time. At 1:30 P.M. local time the core was uncovered. It was recovered after 45 minutes but there was a direct release path (e.g. PORV) to the containment.

We will first run this problem with none of the active reduction mechanisms operating and with a 4-inch diameter opening in the containment with the containment at the maximum design pressure of 15 lb/in^2 . Then we will change release pathways to reduce the size of the release. It is assumed the opening will remain open for 3 hours (until 16:30) and then be completely closed. (Hint: you need to set the diameter to zero.)

Run the following six cases and fill in the source term and doses in the table.

Case 1: Initial case described above (with sprays off, ice bed exhausted, and fans off)

Case 2: Reduce the containment pressure by a factor of 3; from 15 lb/in^2 down to 5 lb/in^2.

Case 3: Reduce the hole size to 2 inch diameter.

Case 4: Assume that the ice bed is not exhausted

Case 5: Turn on the recirculation fans

Case 6: Turn on the sprays

Use the predefined **Standard Meteorology** dataset.

	Case 1	Case 2	Case 3	Case 4	Case 5	Case 6
Sprays	Off	Off	Off	Off	Off	On
Recirculation fans	Off	Off	Off	Off	On	On
Ice bed exhausted	Yes	Yes	Yes	No	No	No
Hole size @ 13:30	4" dia	4" dia	2" dia	2" dia	2" dia	2" dia
Containment pressure @ 13:30	15 lb/in^2	5 lb/in^2	5 lb/in^2	5 lb/in^2	5 lb/in^2	5 lb/in^2
Total source term (Ci)						
TEDE at 1 mile (rem)						

Event Type   Nuclear Power Plant

Event Location   Catawba - Unit 1

Leave the average reactor power and the average fuel burnup at the default values.

Source Term   Time Core is Uncovered

Reactor shutdown: **10:45**

Core uncovered: **Yes, at 13:30**

Core recovered: **Yes, at 14:15**

Release Path   **Containment Leakage / Failure**

Release point characterization: **Not an isolated stack**

Release height: **0.0 m**

Consider building wake effects: **Yes**.

Start of release to containment: **13:30**

> Rationale: The problem specified no holdup in containment. Set the time to match the core uncovered time (13:30).

Leak rate to atmosphere described by: **Containment Pressure / Hole size**

Leak events:

<date>	13:30	Sprays	Off
<date>	13:30	Recirc fans	Off
<date>	13:30	Ice bed exhausted:	Yes
<date>	13:30	Leak rate	15 lb/in², 4 inch dia
<date>	16:30	Leak rate	15 lb/in², 0 inch dia (terminates the release)

> Rationale: The leak rate event at 16:30 is added to close the major release path.

The leak events shown above are for Case 1. Modify the leak events as needed for each of the other cases (2 - 6).

Meteorology   Data set type: **Predefined data (non site specific)**
Data set: **Standard Meteorology**

Calculation Options   Distance of calculation: **Close-in only**.

End calculations at: **Start of release plus 6 hours**

Results

	Case 1	Case 2	Case 3	Case 4	Case 5	Case 6
Sprays	Off	Off	Off	Off	Off	On
Recirculation fans	Off	Off	Off	Off	On	On
Ice bed exhausted	Yes	Yes	Yes	No	No	No
Hole size	4" dia	4" dia	2"dia	2" dia	2" dia	2" dia
Containment pressure	15 lb/in^2	5 lb/in^2	5 lb/in^2	5 lb/in^2	5 lb/in^2	5 lb/in^2
Total source term (Ci)	3.2E+07	2.4E+07	6.5E+06	5.7E+06	5.3E+06	4.9E+06
TEDE at 1 mile (rem)	270	200	53	27	14	2.9

19 Monitored Mixtures

To show how to determine projected doses based on effluent monitor readings.

Discussion

When a nuclear power plant effluent monitor detects a release, the release is likely to be a mixture of many radionuclides (fission products) rather than just a single radionuclide. However, the effluent monitor cannot identity the specific radionuclides present. Instead, the effluent monitor will usually be able to provide only the noble gas activity release rate, the radioiodine activity release rate, and sometimes a particulate activity release rate.

The effluent monitor may provide a reading in terms of some number of counts per second. The counts per second reading cannot be used by RASCAL because RASCAL does not know the efficiency of the effluent monitor. Therefore, RASCAL cannot convert the count rate in terms of counts per second to a concentration in terms of curies per second. However, every plant has a conversion factor that will permit the conversion of counts per second to curies per second.

Sometimes, the plant may use this conversion factor and report the release rates to you directly in terms of curies/second. Other times it may not be easy to obtain this conversion factor from the plant because the factor has been buried in the plant's dose projection software, and the users of the software may enter the monitor reading in terms of counts per second without knowing the value of the conversion factor that is being used. Nevertheless, the plant has this conversion factor, and you should be able to get it if you are persistent. Once you have the conversion factor, you can convert counts per second to curies per second.

If the plant provides you with an effluent concentration in terms of curies per unit volume and the effluent flow rate, you can multiply these two numbers together to determine the effluent activity release rate in terms of curies/second.

Monitored releases will be filtered so they should be mostly noble gases but with a small proportion of iodines and particulates. RASCAL estimates the composition of the noble gas component of the release by starting with the equilibrium proportions of each noble gas radionuclide that will be present in the reactor core while it is operating. Then the proportions of each that will be present at the start of the release will be determined by multiplying the initial proportion by the radiological decay factor and renormalizing the proportions to the time of the release start. The same is done for radioiodines.

For particulates a different technique is used because there is really no practical way to determine the composition of the particulate mixture. Particulates are assumed to be entirely composed of cesium iodide (Cs-137 + I-131). This is a very conservative assumption. It assumes that particulates are entirely made up of radionuclides that have relatively long half lives and are biologically among the most hazardous. However, the overestimate of the dose for particulates in not expected to have a significant impact on protective action decisions because the particulate release rate should be very low.

Problem

A malfunction at the North Anna, Unit 2 nuclear power plant caused the plant to shut down at 3:50 P.M. Shortly thereafter (4:00 P.M.), an effluent release through a monitored pathway was detected.

The effluent release rate was reported to be 95 Ci/s for noble gases, 1.2 Ci/s for radioiodines, and 0.03 Ci/s for particulates. The plant's technical specifications state that the release duration must be limited to no more than 30 minutes.

Determine the projected TEDE and Thyroid CDE at 1 mile.

Dose type	Distance from release
	1 mile
TEDE (rem)	
Thyroid CDE (rem)	

Inputs

Event Type   Nuclear power plant

Event Location   North Anna - Unit 2
Assume 100% power and default burnup

Source Term   Monitored Releases - Mixtures

Reactor shutdown: **Yes, at 15:50**

Sample taken: **16:00**

Release rates:

Noble gases	**95 Ci/s**
Iodines	**1.2 Ci/s**
Particulates	**0.03 Ci/s**

Release Path   **Direct to Atmosphere**

Release point characterization: **Not an isolated stack**

Release height: **0.0 m**

Consider building wake effects: **Yes**

Start of release to atmosphere: **16:00 (** to correspond to the sample time)

End of release to atmosphere: **Release duration - 00:30**

Meteorology Data set type: **Predefined data (non site specific)**
Data set: **Standard Meteorology**

Calculation Options Calculation distance: **Close-in only**

End calculations at: **Start of release to atmosphere plus, 6 hours**

Results

The following results are available from the **Maximum Dose Values** tab displaying the close-in doses.

Dose type	Distance from release
	1 mile
TEDE (rem)	1.3E-01
Thyroid CDE (rem)	1.2E+00

20 Containment Radiation Monitor

Purpose

To learn how a containment radiation monitor reading can be used to generate a source term.

Discussion

RASCAL can use a reading from a containment radiation monitor to estimate the amount of core damage. For example, RASCAL might calculate that the monitor reading you entered corresponds to 37% core melt. RASCAL will then use 37% core melt to calculate the activity released to the containment based on the core inventory and the core release fractions.

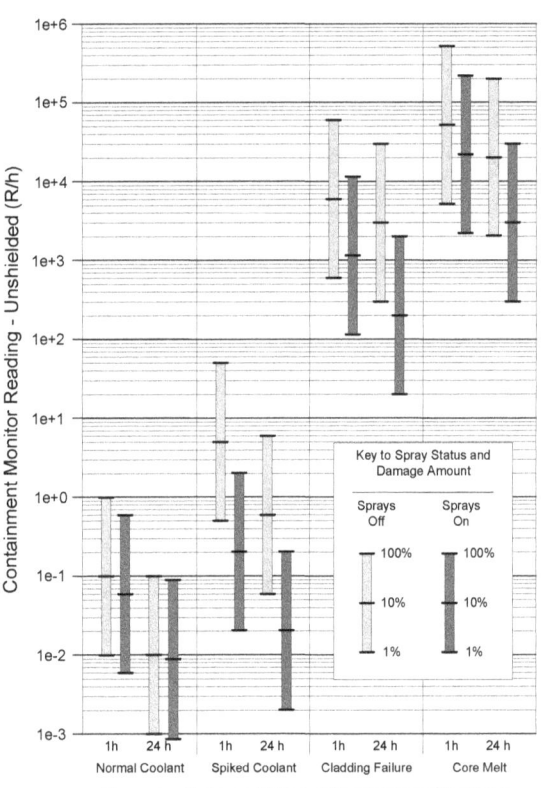

The factors that RASCAL uses to estimate core damage consider whether the sprays are on or not. If the sprays are operating, they will wash out most of the particulates so that a given monitor reading would correspond to more core damage than if the sprays were off. Figure 3 shows the containment dome monitor response for a PWR.

Figure 3 PWR containment monitor response

The advantage of this method is that the core damage estimate is based on real measured data instead of a calculation of how much core damage is likely to occur based on how long the core is uncovered. Used together, the two methods can compliment each other and allow a better understanding of what is really happening.

There are several cautions in using this method. First, the monitor must "see" more than 50% of the area being monitored (see Figures A-3 and A-4 in the RTM). If the monitor does not see this much, do not use this method.

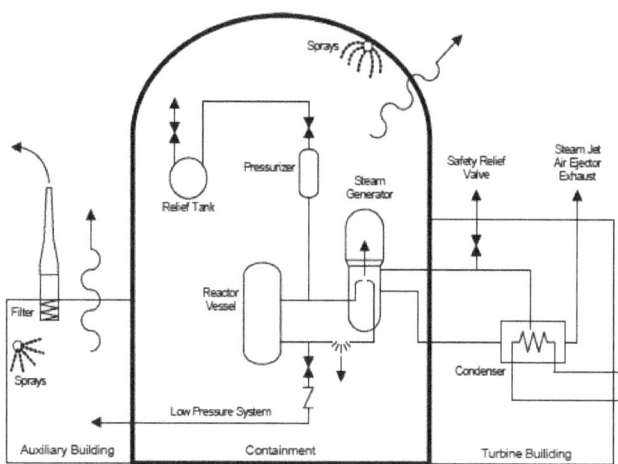

Second, in interpreting the RASCAL results, remember that there will be a time delay before fission products released from the core can affect the monitor reading. When there is a delay in the movement of fission products from the core to the containment or a delay in mixing within the containment, containment radiation monitor readings may start out low and then steadily increase with time. Thus, the maximum reading will give the best estimate of true core damage.

Third, the actual radionuclide mix in the containment may be different from that assumed in the calibration of the monitor. This can be especially significant for high dose rates when a shielded detector is used.

Problem

An accident occurred at Waterford Unit 3. The reactor scrammed at 12:00 P.M. By 12:45 P.M. some of the core was uncovered. The release from the core passed into the containment building. Release from the containment was via design leak rate.

During the course of the event, you receive periodic readings from the containment dome radiation monitor. Perform a RASCAL calculation (with Standard Meteorology) based on the containment radiation monitor readings in the table below:

Time	PWR containment monitor reading, R/h
12:45 P.M.	14
1:30 P.M.	50,000
3:00 P.M.	300,000

Look at the **Source Term Summary** tab and fill in the table below.

Containment radiation monitor reading	Indicated core damage
12:45 P.M.	
1:30 P.M.	
3:00 P.M.	

Inputs

Event Type   Nuclear power plant

Event Location Waterford - Unit 3

Source Term   Containment Radiation Monitor

Reactor shutdown: **12:00**

Monitor location: **PWR Containment Dome** (not user selectable)

Containment radiation monitor readings (R/h):
<date>	12:45	14
<date>	13:30	50000
<date>	15:00	300000

Release Path	**Containment Leakage / Failure**

Release point characterization: **Not an isolated stack**

Release height: **0.0 m**

Consider building wake effects: **Yes**

Start of release to containment: **12:45** to correspond to first monitor reading

Leak rate events:

<date>	12:45	Leak rate	Design
<date>	12:45	Sprays	Off

Meteorology	Data set type: **Predefined data (non site specific)**
	Data set: **Standard Meteorology**

Calculation Options	Calculation distance: **Close-in only**
	End calculations at: **Start of release to atmosphere plus, 6 hours**

Results

Examine the **Source Term Summary** to see what amount of core damage was estimated based on the radiation monitor readings.

Containment radiation monitor reading	Indicated core damage
12:45 P.M.	1.92 E-02% cladding failure
1:30 P.M.	7.2% core melt
3:00 P.M.	43.2% core melt

21 Containment Bypass

Purpose

To learn how to estimate projected doses when there is a release of coolant and when the release pathway bypasses the containment.

Discussion

We have to distinguish between "source term" and "release pathway." A "bypass accident" means we have specified the release pathway. The release pathway for a bypass accident is one in which coolant escapes to the atmosphere without going through the containment.

There are three source term types that can be used for coolant releases. The "time core is uncovered" source term type can estimate the activity entering the coolant if the core is uncovered. The "ultimate core damage state" source term type can estimate coolant activity for normal coolant, for spiked coolant, and for accidents in which there is cladding damage. The "coolant sample" source term type can estimate coolant activity based on the sample data.

Problem

The Peach Bottom Unit 2 nuclear power plant experienced a leak in a containment penetration pipe that could not be isolated (closed). The resulting drop in reactor coolant system pressure and water level caused the reactor to automatically shut down at 11:45 A.M., but some fuel damage occurred. The indication is that the fuel suffered 4% cladding failure.

The pumps being used to maintain the water levels in the reactor coolant system are injecting 300 gallons/minute.

The leak is into an auxiliary building where the coolant immediately flashes into steam. The steam can escape the building to the atmosphere without any filtering.

Calculate the projected TEDE and thyroid CDE at 2 miles.

Dose type	Distance from release
	2 miles
TEDE (rem)	
Thyroid CDE (rem)	

Inputs

Event Type Nuclear power plant

Event Location Peach Bottom - Unit 2
Assume 100% power and default burnup

Source Term Ultimate Core Damage State

Reactor shutdown: **11:45**

Ultimate core damage state: **Cladding failure - 4 percent**

Completion of cladding failure: **11:45**

Release Path **Bypass secondary containment**

Release point characterization: **Not an isolated stack**

Release height: **0.0 m**

Consider building wake effects: **Yes**

Release events:
\<date\> 11:45	Leak rate	300 gal/min
\<date\> 11:45	Filters	Off

Meteorology Data set type: **Predefined data (non site specific)**
Data set: **Standard Meteorology**

Calculation Options Calculation distance: **Close-in only**

End calculations at: **Start of release to atmosphere plus, 6 hours**

Results

The following results are available from the **Maximum Dose Values** tab displaying the close-in doses.

Dose type	Distance from release
	2 miles
TEDE (rem)	1.0E+01
Thyroid CDE (rem)	2.0E+02

---

22 Steam Generator Tube Rupture with Coolant Release

Purpose

To learn how to determine projected doses due to a steam generator tube rupture and to understand coolant spiking.

Discussion

A steam generator tube rupture (SGTR) in a PWR allows primary system coolant to escape rapidly to the secondary system. The rapid loss of coolant causes the pressure to drop in the primary system, which in turn causes the reactor to automatically "scram" (shutdown).

Coolant "spiking"

The rapid drop in pressure in the primary system causes an increase in the rate at which the radioactive fission products in the fuel rod cladding gap escape to the coolant. The concentration of fission products in the coolant rises rapidly. Originally, it was noticed that radioiodine concentrations rose and the phenomenon was first called "iodine spiking." Later it was found that the concentrations of all radionuclides rose with the possible exception of noble gases. Now we call this phenomenon "coolant spiking" or simply "spiking." RASCAL assumes that the concentrations of only the halogens and alkali metals fission products in the coolant increase by the spiking factor.

The RASCAL user must specify the "spiking factor" using the **Ultimate core damage state** screen. At the start of an accident there will be no way to determine what the actual spiking factor is. After some time, it should be possible to analyze a coolant sample and estimate the spiking factor be comparing its activity to normal coolant activity. Thus, we recommend running RASCAL initially using the spiking factor default value of 100 and then adjusting the spiking factor after a primary coolant sample has been analyzed.

Partitioning

The RASCAL user must specify whether the radionuclides that leak into the steam generator are "partitioned" or "not partitioned." "Partitioning" occurs when the leak is below the water level on the secondary side of the steam generator. The water from the primary system that leaks into the steam generator mixes with the water in the steam generator. When the radionuclides are "partitioned," the concentration of radionuclides in the steam will be less than their concentration in the steam generator water by a partitioning factor." "Not partitioned" occurs when the leak is above the steam generator water level. Then, most of the primary coolant will flash into steam and most of the radionuclides will be available for release.

Partitioning is assumed to have no effect on noble gases. Noble gases in the primary coolant leakage are assumed to immediately be available for release without any reduction or holdup. If the user selects "partitioned," RASCAL reduces the release of all radionuclides that are not noble gases by a factor of 50. If the user selects "not partitioned," RASCAL reduces the release of all radionuclides that are not noble gases by a factor of 2.

How does the RASCAL user know whether to select "partitioned" or "not partitioned"? Although the actual tube break location may be unknown, the steam generator water level relative to the top of the tube bundle should be available. If the water level is near or above the top of the tubes, the user should select "partitioned." If the reactor has a once-through steam generator, RASCAL does not allow "partitioning" because most of the tube length in a once-through steam generator is not covered by water.

Setting the leak rate

The RASCAL user must also specify the rate of primary coolant loss to the secondary system. This can be done in one of two ways.

One method is to specify the actual rate of coolant loss from the ruptured tubes. This loss can be estimated by the amount of water being added to the primary, from whatever source, to try and balance the flow into the secondary system. For a typical steam generator leak, the normal charging system flow may be enough to keep up with the leak. In this case, the difference between the letdown flow rate and the charging flow rate is the rate of leakage. Some plants have an "excess makeup flow rate" meter that displays this difference.

The other method is by specifying the number of steam generator tubes that have ruptured. RASCAL then assumes that the rate of primary coolant loss is the number of tubes that have failed multiplied by 500 gpm/tube. It is reasonable to select one tube unless the coolant makeup rate is so large that it indicates that more than one tube might be ruptured.

If the tube leak exceeds the normal charging capacity, then safety injection may actuate when the pressurizer pressure decreases. The operators can estimate the coolant makeup rate from the safety injection flow meter. Note that there was an event at a plant a few years ago, in which the operator detected the leak and started a second charging pump. This prevented the automatic actuation of safety injection and the reactor trip as the two pumps kept up with the leakage.

The bottom line is to enter the rate at which water is being added to the primary coolant system, from whatever source. The program labels it "charging flow" but it might be more aptly labeled "makeup flow".

Release pathways

The RASCAL user must also specify whether the release to the atmosphere is through the steam jet air ejector or the safety relief valves on the secondary system that are used to relieve high pressure. Usually the release will be through the steam jet air ejector.

The primary-to-secondary leakage may cause pressures to briefly exceed the safety relief valve set-point, but the main steam dump valves should quickly relieve the pressure. The response of the steam dump valves may not be quite fast enough to handle a load rejection without a little steam release through the safety relief valves, but this release is generally less than a minute, which is negligible.

When the release is through the steam jet air ejector, RASCAL assumes a reduction factor of 20 for non-noble gases. When the release is through the safety relief valves, there is no reduction. A release through the steam jet air ejector is likely to be a monitored release. In that case, it would be useful to do another RASCAL run using the "Monitored Releases - Mixtures" source term option.

If the main condenser is not available, for example, because of loss of offsite power, the secondary system pressure will increase until the safety relief valves open. Thus, if the main condenser is not available, the RASCAL user should assume the release is through the high-pressure safety relief valves.

As time progresses, the pressure difference between the primary and secondary will decrease and the movement of fission products to the secondary will stop after several hours. Thus, the release is essentially self-terminating.

Problem

The Crystal River Unit 3 nuclear power plant experienced a sudden drop in primary system pressure and a sudden rise in secondary pressure at 12:36 A.M., shortly after midnight. The reactor immediately scrammed (shut down) in response to the sudden drop in primary system pressure.

The operators assumed that a steam generator tube rupture had occurred. The operators estimated that the makeup flow (including safety injection) was about 500 gpm. Assume that the sudden drop in primary system pressure will cause coolant spiking. Use the default spiking factor of 100.

The increase in pressure on the secondary side caused the high-pressure safety relief valves to open briefly, but after that the release was through the steam jet air ejector.

The release point is 30 meters above ground level.

Determine the TEDE and the thyroid CDE at 1 mile.

Dose type	Distance from release
	1 mile
TEDE (rem)	
Thyroid CDE (rem)	

Inputs

Event Type   Nuclear Power Plant

Event Location   Crystal River - Unit 3

Leave the average reactor power and the average fuel burnup at the default values.

Source Term   **Ultimate Core Damage State**
(because only this source term type calculates coolant spiking)

Reactor shutdown: **00:36**

Core damage endpoint: **Increased fuel pin leakage, with coolant contamination spike by factor of 100**

Time of increased fuel pin leakage: **00:36**

Release Path   **Steam Generator Tube Rupture**

Release point: **Steam jet air ejector**

Effective release height: **30.0 m**

> Rationale: This is assumed to be the height of the steam jet air ejector above ground level.

Consider building wake effects: **Yes**

SG water mass: **9.3E+04 lb**

Steaming rate: **7.50E+04 lb/h**

Release events:

<date>	00:36	SG Condition	Not partitioned
<date>	00:36	Leak rate into SG	500 gal/min

Meteorology Data set type: **Predefined data (non site specific)**
Data set: **Standard Meteorology**

Calculation Options You are asked for doses at 1 mile so you could set the distance of calculation to **Close-in only**.

End calculations at: **Start of release plus 6 hours**

Results

Dose type	Distance from release
	1 mile
TEDE (rem)	2.0E-03
Thyroid CDE (rem)	1.5E-02

Conclusion

Offsite doses from a steam generator tube rupture accident are low compared to offsite doses from core damage accidents. A steam generator tube rupture accident without core damage releases only coolant activity. The curie content in coolant without core damage is low compared to the activities that might be present if there were core damage.

23 Coolant Sample

Purpose

To show how to use radionuclide activities measured in a coolant sample to determine the source term.

Discussion

The coolant sample source term option can be used only when coolant is being released directly to the environment. Thus, this source term option can be used for containment bypass release paths and steam generator tube rupture release paths.

The analysis of coolant activities will normally take a couple of hours to complete. Thus, the coolant sample source term option cannot be run early in the accident sequence. Since our objective is to make protective action decisions as early as possible, the coolant sample source term option is not likely to be available in time to be used for making protective action decisions. Other source term options will have to be used instead. However, the coolant sample source term option can be used as a way to improve the accuracy of previous estimates of release size. But, coolant sampling systems are designed for routine sampling and may not be able to measure the high activity concentrations that might result after significant core damage.

Problem

For the previous SGTR problem, assume that the post-accident sampling system (PASS) collected a coolant sample 15 minutes after reactor shutdown. The activity in the coolant was measured as the equivalent to 311 µCi of I-131 per milliliter (after being corrected for radiological decay back to the time the sample was collected).

Determine the TEDE and the thyroid CDE at 1 mile.

Dose type	Distance from release
	1 mile
TEDE (rem)	
Thyroid CDE (rem)	

Compare the source term results with the previous SGTR problem. What conclusions can be drawn?

Inputs

Event Type   Nuclear Power Plant

Event Location	Crystal River - Unit 3

Leave the average reactor power and the average fuel burnup at the default values.

Source Term

Coolant Sample

Time sample taken: **00:51**   (Shutdown time of 00:36 + 15 minutes)

Sample activity units: **µCi/ml**

Nuclide and activities:
    I-131   3.11E+02

Release Path

Steam Generator Tube Rupture

Release point: **Steam jet air ejector**

Effective release height: **30.0 m**

> Rationale: This is assumed to be the height of the steam jet air ejector above ground level.

Consider building wake effects: **Yes**

SG water mass: **9.3E+04 lb**

Steaming rate: **7.50E+04 lb/h**

Release events:
    <date>   00:51   SG Condition   Not partitioned
    <date>   00:51   Leak rate into SG   500 gal/min

Meteorology

Data set type: **Predefined data (non site specific)**
Data set: **Standard Meteorology**

Calculation Options

Distance of calculation: **Close-in only.**

End calculations at: **Start of release plus 6 hours**

Results

Dose type	Distance from release
	1 mile
TEDE (rem)	2.1E-01
Thyroid CDE (rem)	6.1E+00

24 Ultimate Core Damage State Source Term

Purpose

To learn how to use the Ultimate Core Damage State source term.

Discussion

The Ultimate Core Damage State source term option allows you to specify how much spiked coolant or cladding damage you expect and when that "maximum" amount of damage will have been reached.

In RASCAL 3.0.5, the user can no longer select core melt or vessel melt-through as could be done in previous versions of RASCAL. For accidents proceeding to core melt, the user interface screen tells the user to use the "time core is uncovered" source term type. For accidents with core melt, the timing of the release would not be at all realistic using the ultimate core damage state source term option. In addition, while the user may have a relatively good idea of when core damage may begin, he may have less knowledge of when the maximum damage will occur. For these reasons, it is required that for accidents that are expected to proceed into core melt, the use of the "time core is uncovered" source term type should give more realistic results.

Problem

At 8:30 p.m. the Comanche Peak Unit 1 nuclear power plant reactor scrammed when it experienced a sudden drop in primary system pressure. A sudden rise in containment pressure and containment radiation levels indicate that a leak of coolant to the containment atmosphere had occurred. Failure of a pump supplying coolant caused that water level to drop temporarily causing some fuel rods to overheat.

At 09:00 p.m. a coolant radiation monitor indicated that the cladding on 4% of the fuel rods had been damaged.

Assume that the containment sprays were not activated. There is no reason to believe that the containment will not remained intact. Estimate off-site doses.

Dose type	Distance from release
	0.1 mile
TEDE (rem)	
Thyroid CDE (rem)	

Inputs

Event Type Nuclear Power Plant

Event Location Comanche Peak - Unit 1

Leave the average reactor power and the average fuel burnup at the default values.

Source Term **Ultimate Core Damage State**

> Rationale: This is the appropriate source term type since we have an estimate of the total amount of fuel damage that occurred.

Reactor shutdown: **20:30**

Core damage endpoint: **Cladding failure - 4%**

Time of cladding failure: **21:00**

Release Path **Containment Leakage/Failure**

Release point characterization: **Not an isolated stack**

Effective release height: **0.0 m**

Consider building wake effects: **Yes**

Start of release to containment: **21:00**

Release events:

<date>	21:00	Sprays	Off
<date>	21:00	Leak rate	Design

> Rationale: We assume that the sprays were not activated and with an intact containment the leak rate will be at design.

Meteorology Data set type: **Predefined data (non site specific)**
Data set: **Standard Meteorology**

Calculation Options Distance of calculation: **Close-in only**.

End calculations at: **Start of release plus 6 hours**

Results

Dose type	Distance from release
	0.1 mile
TEDE (rem)	2.4E-02
Thyroid CDE (rem)	4.7E-01

Spent Fuel

There are three possible spent fuel source terms from spent fuel accident types: cold gap, hot gap, and cladding fire. For all three, the inventory available for release is computed the same way. The number of assemblies per core is specific to each plant and is part of the plant parameters in RASCAL. The activity of a full core can be computed from the reactor power and the core inventory adjusted for burnup. The spent fuel source term inventory is then the full core inventory scaled as needed for the portion of a full core represented by the number of spent fuel assemblies involved in the accident. Where the spent fuel amount is specified in batches, a batch is defined as one-third of a core. The burnup value used to adjust the inventory is the average burnup of the fuel in storage not the fuel currently in the reactor core.

Cold gap release

A "cold gap release" occurs when spent fuel elements are mechanically damaged but the temperature is low enough that the cladding does not suffer any thermal damage. Examples include mechanical damage to fuel elements damaged underwater in a fuel pool and mechanical damage to fuel elements in a dry cask storage. Radionuclides in the gap between the cladding and the fuel pellets will be released. The fractions of the inventory in the fuel element that will be released in a "cold gap release" are given in Table 5 for each radionuclide group.

Hot gap release

A "hot gap release" occurs when the fuel cladding ruptures due to heat build up releasing the gap inventory. If the spent fuel elements in a pool become uncovered, their temperature will rise. If the temperature reaches 1200 °F, the fuel cladding will rupture due to the buildup up of internal pressure and will release the radioactivity in the gap between the cladding and the fuel pellets. The fractions of the inventory in the fuel element that will be released in a "hot gap release" are given in Table 5 for each radionuclide group.

Cladding fire

A "cladding fire release" occurs when the fuel elements remain uncooled and the temperature becomes high enough to cause the cladding to burn. The fractions of the inventory in the fuel element that will be released in a "cladding fire release" are given in Table 5 for each radionuclide group.

Table 5 Fuel activity release fractions used in spent fuel accidents

Nuclide group	Cold gap	Hot gap	Cladding fire
Noble gases (Xe, Kr)	0.4	0.4	1
Halogens (I, Br)	0.003	0.03	0.7
Alkali metals (Cs, Rb)	0.003	0.03	0.3
Tellurium group (Te, Sb, Se)	1×10^{-4}	0.001	0.006
Barium, strontium (Ba, Sr)	6×10^{-7}	6×10^{-6}	6×10^{-4}
Noble metals (Ru, Rh, Pd, Mo, Tc, Co)	6×10^{-7}	6×10^{-6}	6×10^{-6}
Cerium group (Ce, Pu, Np)	6×10^{-7}	6×10^{-6}	2×10^{-6}
Lanthanides (La, Zr, Nd, Eu, Nb, Pm, Pr, Sm, Y, Cm, Am)	6×10^{-7}	6×10^{-6}	2×10^{-6}

Reference: from Table 2.1 in NUREG-1887; originally from NUREG/CR-6451

Caution: The release fractions for spent fuel are for oxides, not metal.

RASCAL can analyze three types of spent fuel accidents as seen in the **Source Term Options for Spent Fuel** screen below:

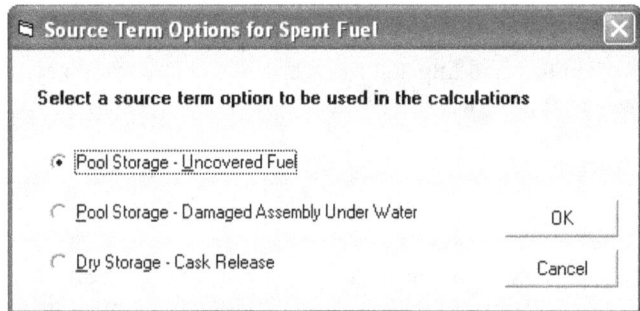

Pool Storage - Uncovered Fuel

RASCAL can analyze spent fuel pool accidents in which spent fuel pool water is lost and the fuel elements become uncovered. When spent fuel pool water is lost, it is assumed that the fuel must be uncovered for more than two hours before sufficient heating can occur to reach 1200 °F, the temperature at which cladding is expected to fail. After two hours, the fuel cladding ruptures due to the build up of internal pressure releasing the hot gap fraction of the fission products (see Table 6). Only fuel that has been irradiated within the past year (for a PWR) or the last 180 days (for a BWR) is assumed to be able to reach 1200 °F after the loss of cooling water. Loss of cooling for older fuel will not result in a release. If the release is filtered, a reduction factor of 0.01 is applied to all radionuclides except noble gases.

Table 6 Spent fuel damage estimates based on storage duration and time fuel is uncovered

Storage duration	Type of release	
	PWR fuel stored less than 1 year - or - BWR fuel stored less than 180 days	PWR fuel stored 1 year or longer - or - BWR fuel stored 180 days or longer
Pool not fully drained or drained for less than 2 hours	Hot gap release	No release
Pool fully drained for 2 hours or longer	Cladding fire release	Hot gap release

Pool Storage - Damaged Assembly Under Water

RASCAL can analyze mechanical damage to spent fuel in a fuel pool in which the fuel remains underwater and adequately cooled. Fuel that is damaged underwater is assumed to remain cold but suffers cladding failure resulting in a cold gap release (see Table 5). Pool scrubbing removes 99% of the non-noble fission products released from the damaged fuel. If the release to the environment is filtered, a further reduction factor of 0.01 is applied to all radionuclides except noble gases.

Dry Storage - Cask Release

RASCAL has the capability to model an accident in which there is damage to the cladding of fuel assemblies stored in a cask and the integrity of the fuel cask is lost. Only one cask can be involved but the user may vary the number of assemblies damaged. If the assemblies are damaged but have not lost cooling for greater than the thermal limit, a cold gap release fraction is used (see Table 5). If cooling has been lost long enough to allow the cladding to melt, then the hot gap release fraction is assumed. No release is assumed if the cladding does not melt or if the cask is engulfed in a fire.

Decay Since Last Irradiation

RASCAL decays the stored fuel based on the specified "time since last irradiation". Both the damaged assembly underwater and the dry storage cask releases assume that all the damaged fuel is of the same age. In these two cases, decay is from the user entered date or for the user entered time in storage. The pool storage with uncovered fuel scenario allows specification of different ages. In this case the decay is done as follows:

Time since last irradiation		Decay applied by the model
PWR	**BWR**	
< 1 year	< 180 days	1 week
1-2 years	180+ days	1 year
> 2 years	NA	2 years

Caution

The current version of the model (v3.0.5) applies only 1 week of decay to any spent fuel in the first age category (< 1 yr PWR, < 180 days BWR). For fuel whose age is toward the higher side of this interval, this can significantly overestimate the activity available for release and thus overestimate the projected dose. This limitation will be changed in a future version of the code.

25 Assembly Damaged Underwater

Purpose

To learn how to assess a spent fuel accident where the damaged fuel remains underwater.

Discussion

See the introduction to spent fuel source terms at the beginning of this section (page 109).

Problem

You are asked to make a dose projection for a fuel handling accident at the Crystal River nuclear power station. You are told that at 12:49 A.M., a fuel assembly that was being removed from the reactor to the fuel pool was smashed against another assembly in the pool and that both were severely damaged. The reactor had been shut down for 6 days. The assembly already in the pool had been there for over a year.

The fuel pool building is 20 meters high and has a filtered exhaust vent near the top. The building ventilation rate is 10 air changes/hour.

Calculate the TEDE and thyroid CDE at 0.1 and 0.2 miles. How many total curies were released? How many radionuclides were represented in the source term?

Fill in the following tables:

Total activity released to atmosphere	curies

Dose type	Distance from release	
	0.1 mi	0.2 mi
TEDE (rem)		
Thyroid CDE (rem)		

Inputs

Event Type Spent Fuel

Event Location Crystal River - Unit 3

Leave the average reactor power and average fuel burnup at the default values because better information was not provided.

Source Term	Pool Storage - Damaged Assembly under Water

Number of fuel assemblies damaged: **1**

> Rationale: Although 2 assemblies were damaged, one had been recently irradiated and the other had been out of the reactor for much longer. The source term will be dominated by the recently removed fuel. If you want to compare the relative source terms you can run RASCAL separately for each assembly.

Age of damaged fuel assemblies: Use the option for **How long in storage** and set to **6 days**.

When did the damage to the fuel occur: **00:49**

Release Path	Release height: **20.0 m**

Release point characterization: **Not an isolated stack**

Consider building wake effects: **Yes**

Start of release to atmosphere: **00:49**.

> Rationale: The release cannot start until the fuel has been damaged.

End of release to atmosphere: **Release duration: 02:00** hh:mm

> Rationale: You might think you should put in 1 hour because at a rate of 100%/hour, all the fission products will have been released. But this is not true. The release rate is a rate, and like with radioactive decay, some always remains within the building. Using 2 hours as the time will get essentially all the fission products out of the building.

Filtered?: **Yes**

Leak rate to atmosphere: **Total failure (100% per hour)**

> Rationale: This is tricky. Since there are 10 air changes per hour, you might assume that you should set the percent volume to 1000% / hour. If you do this, the code will report that the maximum allowed value is 100% / hour. A value of 1000% / hour would be correct if the fission products released from the fuel instantaneously got into the air and were perfectly mixed. But that really takes time. As a practical matter, roughly 100% / hour is about the fastest rate at which the fission products can escape the building.

Meteorology	Data set type: **Predefined data (non site specific)** Data set: **Standard Meteorology**

Calculation Options	Distance of calculation: **Close-in only**.

Rationale: For non-reactor accidents there is rarely a need to calculate beyond 2 miles, and the code will complete the calculations faster.

End calculations at: **Start of release plus 4 hours**.

Rationale: You want enough time for all the material to be released.

Results

The following results can be compiled from the source term summary and maximum values tables after the completion of the calculations:

Total activity released to atmosphere	1.7E+05 curies

Dose type	Distance from release	
	0.1 mi	0.2 mi
TEDE (rem)	2.5E-02	1.4E-02
Thyroid CDE (rem)	5.7E-03	2.1E-03

26 Pool Drained

Purpose

To learn how to model a spent fuel accident where the storage pool loses water and the fuel is uncovered.

Discussion

See the introduction to spent fuel source terms at the beginning of this section (page 109).

Problem

The plant staff are calling you from San Onofre, Unit 2 because there has been an earthquake in the vicinity. The spent fuel pool has lost much of its water due to a large crack possibly flowing into a sink hole. Due to a malfunctioning pump, it has not been possible to provide enough water to make up for the loss. The water dropped to the top of the fuel at 8:49 A.M., and appears likely to continue dropping. Estimates are that the fuel will be fully uncovered by 11:00 A.M. The pool has high density racking and contains one batch of fuel that was unloaded from the reactor only 2 weeks earlier. (A batch is defined as one-third of a core) Another batch was unloaded about a year before that, and 8 batches have been in the pool for longer than 2 years. The spent fuel building has been severely damaged and is in many places directly open to the atmosphere.

Estimate TEDE and thyroid CDE at 1, 5 and 10 miles and summarize the source term. Fill in the following tables:

Total activity released to atmosphere	curies

Dose type	Distance from release		
	1 mi	5 mi	10 mi
TEDE (rem)			
Thyroid CDE (rem)			

Inputs

Event Type   Spent Fuel

Event Location   San Onofre - Unit 2

Source Term	Pool Storage - Uncovered Fuel
	Density of fuel pool racking: **High**
	Amount of fuel in the pool: Use the **Number of batches** option

      < 1 year: **1 batch**
      1 - 2 years: **1 batch**
      > 2 years: **8 batches**

Fuel uncovered: **08:49**

Pool is totally drained: **Yes, at 11:00**

Fuel is recovered: **No**

Release Path   Release height: **0.0 m.**

Release point characterization: **Not an isolated stack**

Consider building wake effects? **Yes**

Start of release to atmosphere: **10:49**.

> Rationale: The release cannot start until at least 2 hours after the fuel has been uncovered.

End of release to atmosphere: **Release duration: 0 days 02:00 hh:mm**

> Rationale: You might think you should put in 1 hour because at a rate of 100%/hour, all the fission products will have been released. But this is not true. The release rate is a rate, and like with radioactive decay, some always remains within the building. Using 2 hours as the time will get essentially all the material out of the building.

Filtered?: **No**.

> Rationale: Since the problem states that the building is open to the atmosphere any filters are likely being bypassed or having very limited effectiveness.

Leak rate to atmosphere: **Total failure (100% per hour)**

> Rationale: As a practical matter, roughly 100%/hour is about the fastest rate at which the fission products can escape the building. This is a good value to use for catastrophic failure of a building.

Meteorology   Data set type: **Predefined data (non site specific)**
Data set: **Standard Meteorology**

Calculation Options   Distance of calculation: **Close in + to 10 miles**

End calculations at: **Start of release plus 6 hours**

Results

After the run, the results can be summarized as follows:

Total activity released to atmosphere	8.6E+07 curies

Dose type	Distance from release		
	1 mi	5 mi	10 mi
TEDE (rem)	5.2E+03	1.2E+03	4.5E+02
Thyroid CDE (rem)	3.9E+04	8.9E+03	3.5E+03

Both TEDE and thyroid PAGs are exceeded at 10 miles. The model should be run again with a calculation radius of 25 miles. However, don't delay making protective action recommendations waiting for these additional model runs.

27 Dry Cask Rupture

Purpose

To learn how to a spent fuel accident where a dry cask storage unit is ruptured.

Discussion

See the introduction to spent fuel source terms at the beginning of this section (page 109).

The RASCAL database contains a information of a variety of dry storage casks. The user may select from the list of cask types to set the number of assemblies or manually enter a value. Table 7 lists the cask types defined and the number of fuel assemblies stored in each.

Table 7 Spent Fuel Dry Storage Cask Types

Cask name	# fuel assemblies	Cask name	# fuel assemblies
CASTOR V/21	21	NAC-STC	26
CASTOR X/32S	32	NAC-UMS	24
FuelSolutions	21	NUHOMS-24P	24
Hi-Storm 100	24	NUHOMS-32P	32
Holtec Hi-Star 100	24	SNC Transtor	24
NAC-C28 S/T	28	TN-24	24
NAC-I28 S/T	28	TN-32	32
NAC-MPC 24 GTCC	24	TN-40	40
NAC-MPC 36 Yankee	36	VSC-24	24
NAC-S/T	26		

Problem

In the late night hours and under cover of a heavy snowstorm, a terrorist entered the plant grounds of the Prairie Island Nuclear Power Station. Unable to gain access to the reactor buildings, he has barricaded himself in the dry cask storage yard. Law enforcement officials are in contact with him and believe he has attached some explosive device to one of the casks.

The current local time is 4:15 A.M. You are asked to make an assessment of the potential consequences of the explosive breaching of a single cask. The cask involved has been identified as a Castor V/21. This is a vertical metal cask on a concrete pad. The experts predict that the explosion will partially rupture the cask and estimate that roughly three-quarters of the assemblies within will be damaged by the shock. The snowstorm has stopped and current winds are reported as light.

Make your calculations and then fill in the tables below. Use an appropriate predefined meteorological data set other than the standard one.

Total activity released to atmosphere	curies

Dose type	Distance from release	
	0.1 mi	0.2 mi
TEDE (rem)		
Thyroid CDE (rem)		

Inputs

Event Type Spent Fuel

Event Location Prairie Island - Unit 1 (could use either unit as fuel storage area serves both and as long as the fuel burnup value is appropriate)

Source Term Dry Storage - Cask Release

Type of cask: **Known: CASTOR V/21**
This sets the number of PWR assemblies in the cask to 21.

Age of damaged fuel assemblies: **How long in storage: 5 years**

> Rationale: This is unknown at this time but the worst case would be the minimum cooling time value of 5 years. This information would be available from technical staff.

Type of event: **Major damage** with **75%** fuel elements damaged

Release Path Release point characterization: **Not an isolated stack**

Release height: **0.0 m**

Consider building wake effects? **Yes**

Start of release to atmosphere: **04:15**

End of release to atmosphere: **Release duration: 02:00 hh:mm**

> Rationale: We are going to assume that the explosion ruptures the cask but has little effect on dispersing the material. Thus, a release duration of 2 hours provides enough time for the material to leave the cask.

Filtered?: **No** (not an option with dry cask storage releases)

Leak rate to atmosphere: **Total failure (100% per hour)**

Rationale: This will be the rate at which the material exits the ruptured cask and enters the atmosphere. The worst case option would be total failure.

Meteorology   Data set type: **Predefined data (non site specific)**
Data set: **Winter - Night - Calm**

Calculation   Calculation distance: **Close-in only**.
Options

Rationale: For non-reactor accidents there is rarely a need to calculate beyond 2 miles, and the code will complete the calculations sooner.

End calculations at: **Start of release to atmosphere plus 4 hours**.

Rationale: You want a little longer than the end of the release so that all the material can exit the cask.

Results

The following results are available after the completion of the calculations. The total activity released is from the top of the source term summary.

Total activity released to atmosphere	3.4E+04 curies

Dose type	Distance from release	
	0.1 mi	0.2 mi
TEDE (rem)	5.3E+00	2.6E+00
Thyroid CDE (rem)	4.0E+00	1.9E+00

Fuel Cycle / UF$_6$ / Criticality

For fuel cycle facilities, RASCAL can calculate doses from: accidental criticalities, UF$_6$ releases, and releases caused by fires and explosions. The following problems show how to use RASCAL to analyze these types of accidents.

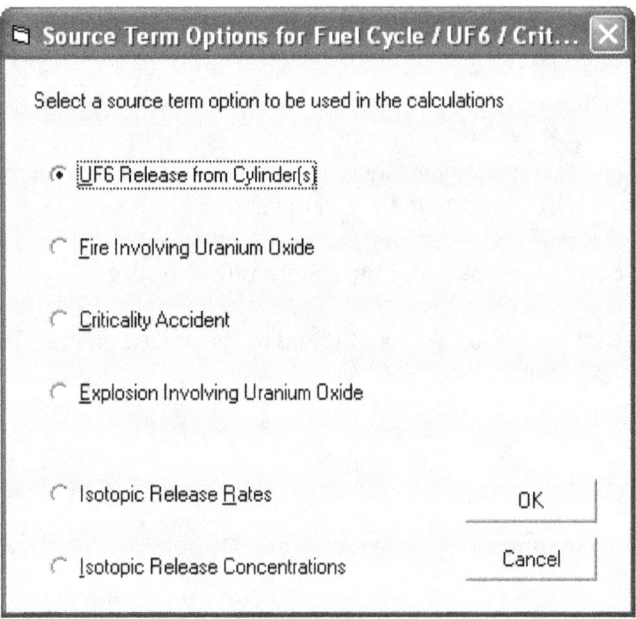

Because we want to focus on the fuel cycle accidents themselves and not meteorology, we will use the predefined **Standard Meteorology** data set for all problems. This is easy to do and eliminates variability caused by differing meteorological conditions.

28 Uranium Process Fire

Purpose

To learn about the options for modeling a fire at a uranium processing facility.

Discussion

Under the Event Type: Fuel Cycle / UF6 / Criticality, it is possible to model the release of uranium oxide resulting from a fire. Five separate fire scenarios are modeled: production process (wet and dry), HEPA filter, incinerator exhaust, waste, and milling. Each scenario may require further refinement of fire conditions by the user. Each combination of fire type and condition has a default airborne release fraction (ARF) and inhalation fraction (IF) as shown in Table 8. These values may be changed by the user.

Table 8 Airborne release fractions and inhalation fractions used in uranium oxide fires

Type of fire	Condition	ARF (Fuel cycle)	ARF (Other radioactive material fires)	IF
Production process	Dry process	0.001	0.001	1
	Wet process	3×10^{-5}	0.001	1
HEPA filter	At high temperature	1×10^{-4}	0.001	1
	Failure	1	NA	1
Incinerator exhaust		0.4	NA	1
Waste fire	Solid packaged in drums	5×10^{-4}	NA	1
	Solid loosely packed	0.05	0.001	1
	Combustible liquid	0.03	0.005	1
	Non-combustible liquid	0.002	0.001	1
Uranium mill	Drum in a fire	0.001	0.001	1
	Solvent extraction	0.03	0.005	1

Reference: from Table 3.7 of NUREG-1887 taken from *Airborne Release Fractions/Rates and Respirable Fractions for Non-Reactor Nuclear Facilities: DOE Handbook*. Vol. 1, 1994

It is also possible to model uranium release from fires using the Event Type: Other Radioactive Material Releases. The default airborne release fraction (ARF) for uranium is 0.001 based on the most commonly encountered forms of the material. However, a table gives alternate airborne release fractions based on the physical form of the material. For comparison of the two approaches, Table 8 also shows ARFs based on that table. In general, the two sets of ARFs are reasonably consistent considering that they came from two different sources and are based on different scenario descriptions and experimental data.

The differences do indicate that some judgement should be used when selecting ARFs. For fuel processing facilities, it is probably more appropriate to use the fire ARFs in the Event Type: Fuel Cycle / UF6 / Criticality than those in the Event Type: Other Radioactive Material Releases. The ARFs used for Fuel Cycle fires were generated specifically for uranium oxide processing. The ARFs for Other Radioactive Materials were generated on a more generic basis.

The source term is calculated using the airborne release fraction and the user specified material at risk (mass of uranium oxide). The mass of uranium oxide is converted to a mass of uranium by multiplying by 0.88. Then the mass of uranium is converted to an activity based on the user specified enrichment value. The inhalation fraction is used during the dose calculations to reduce the amount of material inhaled and contributing to the inhalation dose.

Problem

A fire occurred at 9:47 A.M. in the powder preparation area of the UO_2 Building at the AREVA NP Richland facility. The fire lasted for 20 minutes and breached the building containment. Approximately 12,000 kg of UO_2/U_3O_8 powder (4% enriched) is located in the powder preparation area. All of the airborne material is assumed to be released through the airlock doors.

Fill in the following table:

TEDE at 0.1 mile	
Predominant dose pathway	

Inputs

Event Type   Fuel Cycle / UF6 / Criticality

Event Location   AREVA NP Richland

Source Term   Fire Involving Uranium Oxide

Type of fire: **Production process**

Uranium enrichment level: **4%**

Production process: **Dry Process**

Material at risk: **12,000 kg.**

Airborne release fraction: **0.001** (default)

Inhalation fraction: **1.0** (default)

Rationale: It would also be reasonable to lower the ARF. Experiments involving surface contamination on flammable material (wood) had ARFs of roughly

0.001. When the mass of the uranium is large relative to the mass of the flammable material, lower ARFs can be expected. For this problem an ARF of 10^{-4} or even 10^{-5} might not be unreasonable.

Release Path

Release point characterization: **Not an isolated stack**

Release height: **0.0 m**

Consider building wake effects? **Yes**

Rationale: We have treated this plume as though it will be entrained in the building wake and not rise due to buoyancy which might occur with a heated plume release from a fire. If observers from the site reported that the plume was rising and not being entrained in the building wake, it would be reasonable to select isolated stack and use the release height as the plume rise height estimated by the observers. Building wake would not be used in this case.

Pathway condition: **Inside building** , Pathway filtered: **No**

Rationale: If you expected that there would be significant deposition of the uranium oxide in the building you could decrease the available amount by assuming that the release was filtered.

Release timings
Start of release to atmosphere: **09:47**
End of release to atmosphere: **Duration 0 days 01:00 hh:mm**

Rationale: The release time affects only the rate at which material escapes, not the magnitude. Increasing the release time will leave the release magnitude unchanged, but will decrease the release rate. Remember not to end the calculations before the release has been completed.

Meteorology

Data set type: **Predefined data (non site specific)**

Data set: **Standard Meteorology**

Calculation Options

Distance of Calculation: **Close-in only**.

End calculations at: **Start of release plus 1 hour**.

Rationale: We want to make sure the calculations run for the entire duration of the release. The close-in plume model transports material instantaneously so we don't need to add extra time for transport.

Results

TEDE at 0.1 mile	6.5E-02 rem
Predominant dose pathway	inhalation CEDE

29 Explosion Involving Uranium Oxide

Purpose

To learn about the options for modeling an explosion involving uranium oxide.

Discussion

RASCAL can model the release of uranium oxide from an explosion. First the user must characterize the explosion as one of the following:

– detonation; a shattering such as from a high explosive
– deflagration; enveloping with no shattering such as from a propane explosion
– venting; a result of over-pressurization such as a container rupture from high internal pressure

Next, the form of the uranium oxide material must be specified as liquid, solid, powder, or surface contamination. Each combination of type and form has a default airborne release fraction (ARF) and inhalation fraction (IF) as shown in Table 9. These values may be changed by the user.

Table 9 Airborne release fractions and inhalation fractions for explosions involving U oxide

Explosion characteristics	Material form	ARF	IF
Detonation (high explosive)	Liquid	1	1
	Solid	1	0.2
	Powder	1	0.2
	Surface contamination	0.001	1
Deflagration (fire)	Liquid	1×10^{-6}	1
	Solid	0	0
	Powder	0.005	0.3
	Surface contamination	0.001	1
Venting (sudden pressure change)	Liquid	0.002	1
	Solid	0	0
	Powder	0.1	0.7
	Surface contamination	0.001	1

Reference: from Table 3.8 of NUREG-1887, corrected and taken from *Airborne Release Fractions/Rates and Respirable Fractions for Non-Reactor Nuclear Facilities: DOE Handbook*. Vol. 1, 1994

The source term is calculated using the airborne release fraction and the user specified mass of material at risk (mass of uranium oxide). The mass of uranium oxide is converted to a mass of uranium by multiplying by 0.88. Then the mass of uranium is converted to an activity based on the user specified enrichment value. The inhalation fraction is used in the calculation of inhalation dose.

Problem

You are asked to respond to an accident in the Research and Test Reactor Fuel Element area of Bay 15 at BWX Technologies. At 3:35 P.M. local time, a boxline containing 3.6 kg of U_3O_8 explodes due to a loss of the inert atmosphere and a combination of aluminum and uranium which ignited. The explosion has destroyed the boxline and breached the roof of the building (estimated to be 12 meters above the ground level). The enrichment of the material is 93%.

How many total curies of activity were released to the atmosphere? Estimate the possible doses at the site boundary (540 m away) and to the nearest resident (1100 m away).

Inputs

Event Type   Fuel Cycle / UF6 / Criticality

Event Location   BWX Technologies

Source Term   Explosion Involving Uranium Oxide

Characterize the explosion: **Deflagration**

> Rationale: This is the appropriate explosion type for the ignition of aluminum and uranium.

Material form: **Powder**

Material at risk: **3.6 kg**

ARF and IF: **5.00E-03** and **3.00E-01**

> Rationale: Leave at defaults since have no reason to change.

Uranium enrichment: **93%**

Release Path   Pathway description: **Breached roof**

Release point characterization: **Not an isolated stack**

Release height: **12 m**

Consider building wake effects: **Yes**

Pathway condition: **Inside building, Pathway filtered: No**

Rationale: The explosion occurred inside the building but since the roof has been breached assume that the release will not be filtered or sprayed. A 50% reduction will be applied due to the retention of material within the structure. Thus, you cannot just set the release as being outside to get the same effect.

Release timings:

Start of release to atmosphere: **15:35**

End of release to atmosphere: **15:50**

Rationale: Start was at 15:35. Since it was an explosion, assume that the energetic release got everything out in 15 minutes. Either set the end time to 15:50 or the duration to 00:15.

Meteorology	Data set type: **Predefined data (non site specific)** Data set: **Standard Meteorology**
Calculation Options	Select **Close-in only**. Then, select to use **User defined** close-in distances. Click the **Set Close Distances** button. Then, edit the settings to change from units of miles to kilometers. Also, change two of the default distances to match the two distances of specific interest. For example, change 0.5 to 0.54 and 1.0 to 1.1 km. End calculations at: **Start of release to atmosphere + 1 hour**

Results

The following results are available after the completion of the calculations:

Total source term (curies)	8.7E-04

Dose type	Distance from release	
	0.54 km	1.1 km
TEDE (rem)	2.4E-04	1.3E-04

You will note that the doses are very small - so small that some are not shown on the **Maximum Dose Values** table. However, by selecting **Detailed Results** and then selecting the option of a **Numeric Table** for the Close-in Calculations these lower doses can be seen. Doses are small because the size of the release was very small. There was relatively little uranium involved and uranium has a low specific activity. Also, since uranium is not volatile, relatively little becomes airborne (recall the airborne release fraction was 0.005).

30 UF₆ Cylinder Rupture

Purpose

To learn about the result types available from the UF_6 plume model and how to model a release from a UF_6 cylinder.

Discussion

RASCAL can model the release of UF_6 from a cylinder (or other container). The user must specify the amount of UF_6 contained or the number and type of cylinders. The user must also specify a UF_6 release rate and the fraction ultimately released. The exception to this is for liquid UF_6 in a cylinder with a damaged valve. In this case, RASCAL estimates the fraction released based upon the position of the valve.

RASCAL uses a special model when treating the transport and diffusion of a UF_6 release. The model considers both the buoyancy of the plume due to the heat generated by it's reaction with water vapor and the slumping of the plume caused by its high density. The model is described in detail in Chapter 5 of the RASCAL technical manual.

For UF_6 releases, only one meteorological observation can be entered. In other words, the weather cannot change during the release. This is only true for UF_6 releases and not for any other types of releases.

The model calculates 7 different types of results for UF_6 releases. The types are described below:

HF Concentration - 1-hour equivalent

The damage done by exposure to HF is <u>not</u> strictly proportional to its concentration times the time of exposure as it is for many other toxic gases. Instead, exposure to high concentrations of HF for short times has more effect than the same exposure spread out over a longer time. The HF concentration 1-hour equivalent provides a means of comparing the health effects of exposures by using a common time of exposure. (Reference: S. McGuire, NUREG-1391, *Chemical Toxicity of Uranium Hexafluoride Compared to Acute Effects of Radiation*, 1991). The 1-hour equivalent concentration $C_{1\,hr}$ is a measure of the degree of acute lung edema that may result and cause death. It is calculated in terms of the concentration C during the time of exposure t in hours:

$$C_{1\,hr} = C \times t^{0.5}$$

The implications of this relationship are that a doubling of concentration would require 1/4 the exposure time to maintain the same toxic load level. Or, to double the exposure duration would require the concentration to be about 71% of the original level to maintain effect.

Table 10 lists the health effects and ERPG levels for HF 1-hour equivalent concentrations.

Table 10 Health effects due to HF exposure

Health effect due to HF exposure	HF concentration - 1 hr equivalent
50% fatality without emergency medical treatment	1070 mg/m^3 (1264 ppm)
ERPG-3: American Industrial Hygiene Association Emergency Response Planning Guideline -3: The maximum airborne concentration below which it is believed that nearly all individuals could be exposed for up to one hour without experiencing or developing life-threatening health effects.	41 mg/m^3 (48 ppm)
ERPG-2: The maximum airborne concentration below which it is believed that nearly all individuals could be exposed for up to one hour without experiencing or developing irreversible or other serious health effects or symptoms which could impair an individual's ability to take protective action.	16.4 mg/m^3 (19 ppm)
ERPG-1: The maximum airborne concentration below which it is believed that nearly all individuals could be exposed for up to one hour without experiencing other than mild, transient adverse health effects or without perceiving a clearly defined objectionable odor.	4.1 mg/m^3 (4.8 ppm)

HF Concentration - Average

The HF concentration averaged over the duration of the exposure.

HF Deposition

The HF concentration (g/m²) deposited on the ground. This result provides a measure of the potential hazard from contact with exposed surfaces. It may be used to reconstruct the path of the plume and the HF concentration in the plume. A dry deposition velocity of 0.3 cm/s is used in the calculations.

Uranium Exposure

The uranium exposure (g-s/m³) equal to concentration times the time of exposure.

Uranium Intake

The uranium intake (mg) by inhalation. Uranium intake provides the best measure of potential acute kidney damage due to heavy metal poisoning as shown in Table 11. There are no known long term chemical injuries from uranium intakes that are sub-lethal.

Table 11 Health effects due to uranium intake by inhalation

U intake by inhalation	Health effect
below 5 mg	none
8 mg	threshold for clinically observable transient chemical changes in urine indicating renal distress
50 mg	threshold for permanent renal damage (speculative - never observed in humans)
230 mg	50% fatality without medical treatment

(From: S. McGuire, NUREG-1391).

Uranium CEDE

The committed effective dose equivalent (CEDE) from inhaled uranium (rem) provides the radiation dose to people in the plume. It is a measure of the long-term risk due to radiation exposure. There is essentially no external radiation exposure from airborne uranium so the CEDE will also be the TEDE. The dose conversion factor is 2.73 rem/μCi intake. This value is for soluble uranium.

Uranium Deposition

The uranium concentration (g/m²) deposited on the ground. This result provides a measure of the potential hazard from contact with exposed surfaces and from resuspension of uranium. It may be used to reconstruct the path of the plume and the uranium concentration in the plume. A dry deposition velocity of 0.3 cm/s is used in the calculations.

Problem

At 9:30 A.M. workers at AREVA NP Richland were moving a Model 30 cylinder full of hot, liquid, 4% enriched UF_6 to an outdoor storage area. The cylinder was dropped resulting in the rupture of the cylinder. Evaluate the impact of this release on a person standing on the plume centerline, at a distance of 0.2 miles.

Using the results shown in the maximum dose values table and Tables 10 and 11, fill in the table below:

Result type	Result at 0.2 miles	
	Value and units	Impact
HF conc 1 hr equivalent		
Uranium intake		
Radiation dose (U CEDE)		

Event Type   Fuel Cycle / UF6 / Criticality

Event Location   AREVA NP Richland

Source Term   UF6 Release From Cylinder

Amount of UF_6 involved: Since the cylinder type is known, leave the option selected for **Type and number of cylinders**. Then, set the number of 2½ ton cylinders to **1**.

Form of UF_6: **Liquid**

Cylinder damage type: **Cylinder rupture**

Release fraction: **0.65**

Release rate: **32 kg/s**

Rationale: Given no other information, leave the release fraction and release rates as defaults.

Start of release: **09:30**

Uranium enrichment level : **4%**

Release Path   Effective release height: **Fixed at ground level.**

Rationale: There is no option to enter an elevated release as the UF_6 model handles only releases at ground level.

Consider building wake effects: **Yes**

Release pathway to atmosphere: **Direct**

Rationale: The problem description says the cylinder is outside so the pathway should be direct to the atmosphere.

No other information is needed to define the release pathway.

Meteorology   Data set type: **Predefined data (non site specific)**
Data set: **Standard Meteorology**

Calculation Options   For setting the distance of calculation, notice that **Close-in only** is selected and other choices are unavailable. For releases of UF_6, RASCAL automatically switches to the UF_6 plume model. The user may set specific distances at which doses are calculated.

End the calculations at: **Start of the release to atmosphere + 1 hour**

Results

The following values are available in the Maximum Dose Values table:

Result type	Result at 0.2 miles	
	Value and units	Impact
HF conc 1 hr equi.	4.0 ppm	None. Below level where symptoms are observed
Uranium intake	2.0 mg	None. Below threshold for any observable transient effect
Radiation dose (U CEDE)	0.013 rem	None. Below EPA PAGs and public dose limit

31 UF₆ Cascade Release

Purpose

To learn how to model an accident in a cascade at one of the gaseous diffusion plant buildings.

Discussion

The UF_6 cascade release accident type is available only for the Portsmouth and Paducah gaseous diffusion plants. In addition, the cascade release source term option is available only for certain buildings of those facilities. Tables 12 and 13 list the building names and information about the default inventories and release rates.

Table 12 Paducah GDP buildings and default inventory and release rates

Building name	Cells per unit	Number of units	Avg cell inventory (lbs)	Release rate (lbs/sec)
C-331	10	4	4,400	130
C-333	10	6	9,500	130
C-335	10	4	4,600	130
C-337	10	6	8,400	130
C-310	10	1	150	130

Table 13 Portsmouth GDP buildings and default inventory and release rates

Building name	Cells per unit	Number of units	Avg cell inventory (lbs)	Release rate (lbs/sec)
X-326	20	2.5	1,000	130
X-330	10	11	5,000	130
X-333	10	8	5,000	130

Problem

You are called upon to make a dose assessment for an accident at the U.S. Enrichment Paducah gaseous diffusion plant. At 12:45 P.M. a block valve control failure resulted in a pressure increase which ruptured the primary containment of a cell. The problem occurred in the C-333 process building and operators have been unable to isolate the cell. The enrichment level is 5%.

It is a windy, summer afternoon with the process building operating in its summer flow configuration. What are the potential radiation and chemical doses to persons 2 miles downwind of the building?

Fill in the following table with values from the model and impacts taken from Tables 10 and 11.

Result type	Results at 2 miles	
	Numerical value	Impact
Radiation dose (U CEDE)		
Uranium intake		
HF conc 1 hr equi.		

Inputs

Event Type   Fuel Cycle / UF6 / Criticality

Event Location   U.S. Enrichment - Paducah

Source Term   Building: **C-333 Cascade / Process**
UF6 Release From Cascade System

Release described by: **1 cell with 9,500 pounds per cell**

Rationale: We are told that a single cell containment failed. Since we do not have information on the cell inventory, accept the default value.

Start of release: **12:45**

Release rate to building: **130 pounds per second**

Rationale: Since we have no additional information, accept the default value for the release rate.

Release fraction: **1.0**

Rationale: The worst case is for all the material to move from the cascade cell into the building so leave the default value unchanged.

UF_6 Enrichment level: **5%**

Release Path   Cascade building flow configuration: **Summer**

Meteorology   Data set type: **Predefined data (non site specific)**

Data set: **Summer - Afternoon - Windy**

Rationale: Problem states a windy, summer afternoon and no onsite weather data is available at this time.

Calculation Options

There is no option for calculation distance with UF_6 releases. The UF_6 plume model is always selected. You may however adjust the distances of the 8 receptor rings if needed. In this problem, leave the distances as set.

End of calculations: **Start of the release to atmosphere + 2 hours**

Results

Result type	Result at 2 miles	
	Numerical value	Impact
Radiation dose (U CEDE)	2.3E-03 rem	None. Below EPA PAGs and public dose limit.
Uranium intake	0.28 mg	None. Below soluble uranium intake limit of 10 mg in 10CFR Part 20
HF conc 1 hr equi.	0.46 ppm	None. Below ERPG-1

The distance at which 100% of the UF_6 has reacted with the moisture in the atmosphere is 96.6 meters. (This is shown on the Maximum Dose Values tab of the main screen).

Essentially, the radiation dose (U CEDE) is a TEDE since the cloudshine and groundshine doses are effectively zero. Thus, the U CEDE can be compared to the EPA TEDE PAG (see page 39).

32 Criticality

Purpose

To learn how to use RASCAL to estimate consequences from a criticality accident.

Discussion

To estimate doses from a criticality accident, RASCAL first establishes the number of fissions that will occur. The user may either enter the number of fissions directly or may select the type of system being modeled, allowing RASCAL to determine the fission yield. The systems available and their assumed fission yields are shown in Table 14. The total yield assumes multiple bursts over 8 hours.

Table 14 Fission yields used in criticality calculations

System modeled in the scenario	Initial burst yield (fissions)	Total yield over 8 hours (fissions)
Solution <100 gal	1×10^{17}	3×10^{18}
Solution >100 gal	1×10^{18}	3×10^{19}
Liquid/powder	3×10^{20}	3×10^{20}
Liquid/metal pieces	3×10^{18}	1×10^{19}
Solid uranium	3×10^{19}	3×10^{19}
Solid plutonium	1×10^{18}	1×10^{18}
Large storage arrays below prompt critical	Steady state	1×10^{19}
Large storage arrays above prompt critical	3×10^{22}	3×10^{22}

Reference: from Table 3.9 of NUREG-1887 which were taken from NUREG/CR-6410

From the fission yield, RASCAL determines the fission product release to the atmosphere, the atmospheric dispersion of those fission products, and the doses from the plume. RASCAL also calculates the direct gamma and neutron shine dose. The user can reduce the amount of material released to the atmosphere by specifying release fractions for noble gases, iodines, and other radionuclides. Further reductions can be specified by having a filtered release pathway. The user can also specify shielding to reduce the direct shine dose.

Problem

At 1:45 P.M., criticality alarms sound at the AREVA NP Richland facility. The operators determine that a criticality excursion is occurring in a 125 gallon tank holding a solution of enriched UO_2 and water. Subsequently, radiation monitors detect a second burst indicating that the criticality is on-going. The room air is vented through an 18 meter high exhaust stack on the building roof.

1. How many total fissions will occur?

2. Fill in the following table:

Dose (rem)	Distance		
	150 m	300 m	3,000 m
TEDE from plume			
TEDE from criticality shine			
Total TEDE			

3. To what distances are the EPA PAGs exceeded?

4. What early health effects would we expect to a person standing 50 m from the criticality when it occurred?

Inputs

Event Type   Fuel Cycle / UF6 / Criticality

Event Location   AREVA NP Richland

Source Term   Criticality Accident

Start of criticality: **13:45**

Criticality consists of: **Multiple bursts**

Rationale: This an appropriate choice since measurements suggest that the criticality is continuing.

Define fission yield by: **Type of system with Solutions over 100 gal**

Airborne release fractions: **1.0, 0.25, and 0.0005 (defaults)**

Rationale: There is no other information that would cause us to change the default settings.

Shielding thickness: **0.0 for steel, concrete, and water**

Rationale: There is no information to cause a change from the defaults. You could assume no shielding which would be the worst case.

Release Path	Pathway description: **Room air vented through exhaust stack**

Release point characterization: **Not an isolated stack**

Rationale: Even though the description calls for the release to be through an exhaust stack, this is not an isolated stack. For RASCAL and emergency response, a stack should be considered "isolated" if it is more than 2 times the height of a building that is within a distance of 5 stack heights. Short stacks on roofs should not be treated as isolated.

Release height: **18 m**

Consider building wake effects: **Yes**

Release timings:

Start of release to atmosphere: **13:45**

End of release to atmosphere: **Duration 0 days 10:00 hh:mm**

Rationale: The start of the release would be the start of the criticality. For the duration, the worst case would be if the criticality continues for the full 8 hours. The additional 2 hours will allow most fission products to escape the building. Stopping the release early effectively stops the criticality since the material can no longer be released.

Reduction factors: **1.0** for all 3 groups

Rationale: If you thought there would be some plate-out and deposition within the building and exhaust system, you could assume 50% reduction of non-noble gases during passage through the building and the exhaust system.

Leak rate: **100%/h (ordinary ventilation)**

Rationale: Pick the worst case since there is no information about the flow rates of the exhaust fan.

Meteorology	Data set type: **Predefined data (non site specific)**
	Data set: **Standard Meteorology**

Calculation Options	You are asked for doses only to 3,000 meters (~2 miles). Select **Close-in only**.

To get the doses calculated at 150, 300, and 3,000 meters exactly, select the option to use **User defined** close-in distances. Then select the **Set Close Distances** button. In the section to the left labeled "For non-UF6 releases", select the **Kilometer** option and then make the following changes to distances:

Change distance 1 to:　　**0.15**
Change distance 8 to:　　**3.0**

End calculations at: **Start of release plus 10 hours**.

> Rationale: We want to make sure the calculations run for the entire duration of the release. We added 2 hours to allow material to escape the building. That would also allow time for most of the material to reach 2 miles.

Results

1. From the **Criticality shine dose** option on the **Maximum Dose Values** tab, you can see that there were 3.0E+19 fissions during the release period.

2. The close-in dose table of the **Maximum Dose Values** section will show the TEDE from the passing plume. The table will include the distances set earlier before the calculations. Select the option to see the criticality shine doses. Now, to get the distances in meters requires a shift to the metric units. This however, also changes the dose units to Sv. Simply convert the Sv to rem by multiplying by 100 (1 Sv = 100 rem).

   As you can see from the table you filled out, the dose is dominated by the direct neutron shine dose. Dose from the exposure to the plume is negligible by comparison. RASCAL does not provide shine doses beyond 300 meters. To obtain the dose at 3,000 meters, the inverse square law can be used (divide 300 m dose by 10^2). This approximation neglects attenuation in air and will therefore be an overestimate.

Dose (rem)	Distance		
	150 m	300 m	3,000 m
TEDE from plume	0.35	0.15	0.019
TEDE from criticality shine	140	34	0.34
Total TEDE	140	34	0.36

3. If you consider the total dose delivered, the 1-rem PAG is exceeded to roughly 2,000 meters. However, the PAG refers to projected dose that is potentially avoidable if protective actions are taken. Once the first burst has occurred, its shine dose is no longer part of the projected dose. Thus, it could legitimately be subtracted from the projected dose. But doing that would not change the distance significantly.

4. At a distance of 50 meters, the shine TEDE from all the bursts over 8 hours is 1,200 rem. However, it is reasonable to assume that after the initial burst, the individual would leave the area. Return to the source term screen to redefine the criticality accident to consist of a single burst instead of the multiple bursts. In that case, the shine dose from the first burst would be 37 rem neutron and 3.7 rem gamma. No early health effects would be expected from that dose.

Other Radioactive Material Releases

RASCAL can calculate doses from the release of radioactive materials from other types of facilities. Source terms can be produced for radioactive materials involved in fires or by measured or estimated radionuclide release rates or release concentrations.

Five licensed sites are included in the RASCAL facilities database. In addition, the user can describe a site not in the database.

All radioactive material licensees have much smaller radioactive material inventories than reactors. Thus, the size of potential releases is vastly less. In addition, reactor accidents generally take a long time to progress to a release. This is not always true at materials facilities. Thus, the RASCAL operator may be trying to estimate what *did* happen rather than what *will* happen.

33 Radioactive Material in a Fire

Purpose

To learn how to use RASCAL to model the consequences of radioactive material dispersed in a fire.

Discussion

The **Sources and Material in a Fire** source term option is used to model the consequences of radioactive material dispersed as a result of a fire. The user must specify the amount of material at risk and for each radionuclide set an airborne release fraction and inhalation fraction. A default release fraction (see Table 15) is set by the code for each nuclide selected. These release fractions assume the material is on a completely combustible surface. For example, they would be appropriate for materials on a wooden laboratory bench. They would overestimate the airborne release fraction for material sitting on concrete or steel.

Table 15 Fire activity release fractions by element (based on the likely form of the compound)

Element	Activity Release Fraction	Element	Activity Release Fraction	Element	Activity Release Fraction	Element	Activity Release Fraction
H(gas)	0.5	Se	0.01	I	0.5	W	0.01
C	0.01	Kr	1.0	Xe	1.0	Ir	0.001
Na	0.01	Rb	0.01	Cs	0.01	Au	0.01
P	0.5	Sr	0.01	Ba	0.01	Hg	0.01
S	0.5	Y	0.01	La	0.01	Tl	0.01
Cl	0.5	Zr	0.01	Ce	0.01	Pb	0.01
K	0.01	Nb	0.01	Pr	0.01	Bi	0.01
Ca	0.01	Mo	0.01	Pm	0.01	Po	0.01
Sc	0.01	Tc	0.01	Sm	0.01	Ra	0.001
Ti	0.01	Ru	0.1	Eu	0.01	Ac	0.001
V	0.01	Rh	0.01	Gd	0.01	Th	0.001
Cr	0.01	Ag	0.01	Tb	0.01	Pa	0.001
Mn	0.01	Cd	0.01	Ho	0.01	U	0.001
Fe	0.01	In	0.01	Tm	0.01	Np	0.001
Co	0.001	Sn	0.01	Yb	0.01	Pu	0.001
Zn	0.01	Sb	0.01	Hf	0.01	Am	0.001
Ge	0.01	Te	0.01	Ta	0.001		

Reference: NUREG-1140

Guidance is provided in Table 16 for resetting the airborne release fraction if the element is in some other form.

Table 16 Fire activity release fractions by form of compound

Form of compound	Activity release fraction
Noble gas	1.0
Very mobile form	1.0
Volatile or combustible compound	0.5
Carbon	0.01
Nonvolatile compound / dispersible	0.01
Nonvolatile compound / less dispersible	0.001
U and Pu metal	0.001
Nonvolatile in a flammable liquid	0.005
Nonvolatile in a non-flammable liquid	0.001
Nonvolatile solid	0.0001

Reference: NUREG-1140

The release pathway settings control only the start and stop of the release to the atmosphere. It is assumed that all the material will be released from the facility. There are no provisions with this source term option for the application of sprays or filters.

Problem

You are tasked with estimating the doses resulting from a fire-induced release from a laboratory at the E.R. Squibb radiopharmaceutical facility. At 8:00 A.M. this morning, an electrical short apparently ignited some stored materials in the building and the fire has spread. It is reported that the laboratory may contain as much as 100 curies of I-131.

What is the source term? That is, what is the total activity of I-131 released to the atmosphere?

Estimate doses to persons within a 1 mile radius of the fire and then fill in the following table:

Dose type	Distance from release (miles)	
	0.1	1.0
TEDE (rem)		
Thyroid CDE (rem)		

Event Type   Other Radioactive Materials Release

Event Location   E.R. Squibb

Source Term   Sources and Material in a Fire

                        Fire location: **Laboratory**

                        Activity units for material at risk: **Ci**

                        U enrichment level: **any value is OK**

> Rationale: The uranium enrichment value is used only if enriched uranium is added to the list of material at risk.

                        List of material at risk in the fire: **I-131** and **100.0** Ci

> Rationale: When the I-131 is added, the default values for airborne release fraction and inhalation fraction are entered automatically. These could be changed if you had additional information to support different values. For now, leave the ARF and IF for iodine (0.5 and 1.0) unchanged. Later credit could be given for other reductions such as fire suppression sprays or hold up in the structure.

Release Path   Release point characterization: **Not an isolated stack**

                        Release height: **0.0 m**

                        Consider building wake effects: **Yes**

> Rationale: We have treated this plume as though it will be entrained in the building wake and not rise due to buoyancy which might occur with a heated plume release from a fire. If observers from the site reported that the plume was rising and not being entrained in the building wake, it would be reasonable to select isolated stack and use the release height as the plume rise height estimated by the observers. Building wake would not be used in this case.

                        Release timings:
                        Start of release to atmosphere: **08:00**

                        End of release to atmosphere: **Release duration** option and **01:00**

> Rationale: We don't know how long the fire burned. For simplicity, until better information is available, assume the fire burns for one hour.

Meteorology   Data set type: **Predefined data (non site specific)**
                        Data set: **Standard Meteorology**

| Calculation Options | Distance of calculation: **Close-in only** |

End the calculations at: **Start of release to atmosphere + 1 hour**

> Rationale: Because we are using only the close-in model, there is no need to add time for plume transport. The close-in model transports the plume instantaneously.

Results

The reported source term is 50 Ci of I-131. That is what would be expected given the release fraction of 0.50.

The following doses are projected:

Dose type	Distance from release (miles)	
	0.1	1.0
TEDE (rem)	7.0E-02	5.6E-03
Thyroid CDE (rem)	2.1E+00	1.6E-01

The thyroid dose at a tenth of a mile is less than 50% of the PAG of 5 rem. Persons closer may receive a dose which exceeds the PAG. However, to receive this dose the plume could not rise due to buoyancy and an individual would have to remain on the plume centerline (in the heaviest smoke) for the duration of the release. It is more likely that people would move out of heavy smoke.

34 Transportation Accident

Purpose

To learn how to do an assessment for an accident location not in the RASCAL database.

Discussion

A unique feature of a transportation accident is that its location will not be in the RASCAL database. Thus, the RASCAL operator must enter a new site. This will also be true of accidents at smaller licensed facilities which are not in the RASCAL database. Another unique feature is that information from the site is likely to come from emergency responders with only limited knowledge of radioactive materials.

Most transportation accident scenarios will use the **Other Radioactive Material Releases** event type. However, for spent fuel transportation accidents, select the **Spent Fuel** event type instead.

Problem

At 2:00 A.M. a tractor trailer truck jack-knifed in central Pennsylvania near the intersection of I-80 and I-180 when the driver lost control on the icy road. The truck manifest says it was carrying 150,000 Ci of tritium gas and was bound for the Safety Light facility in Bloomsburg, PA. State highway patrol reports that the trailer slid at high speed into a bridge support and split open.

Make a preliminary assessment of the risk to persons in the immediate vicinity of the crash. What doses may the first responders have received? What doses might nearby residents have received?

Dose type	Distance from release (miles)			
	0.1	0.2	0.5	1.0
TEDE (rem)				
Thyroid CDE (rem)				

As you probably understand, RASCAL results will play only a small part in answering those questions. Many other factors must be considered, some technical and some political. Nevertheless, RASCAL results can still be valuable.

Inputs

Event Type   Other Radioactive Material Release

Event Location   Select the option **Describe a Material Site not in the Database** and then fill in the known information about the location.

The exact location may or may not be important depending on how you will handle the meteorology. You can use "generic" weather (as will be described in the Meteorology section) or you may want the best weather data available. Most likely you will make the first RASCAL run with appropriate "generic" weather and then switch to obtaining actual observations as time permits and they are available. Initially, people will want your RASCAL results as soon as possible. Then, when they have your first results, they will start to ask how accurate the results are. At this point, generic weather is unlikely to be acceptable.

A good way to locate nearby weather stations is by knowing the latitude and longitude of the release point. At the NRC, this information should be readily available from a GIS Mapping Operator on the response team. Alternatively, identifying close cities will help.

For this problem, assume that the GIS operator has given you the following site information:

     latitude =   41.0495° N
     longitude = 76.8400° W
     elevation = 153 meters
     time zone = Eastern

Enter the above information to describe the physical location of the release. Note that you will need to enter the longitude as a negative number since you cannot specify 'west'.

Name: **Truck accident I-80 and I-180**

City, county, state: Consulting a road map you see that the town of Milton is close by. Enter the town and state names.

Source Term   **Isotopic Release Rates**

In order to know how to handle the source term you will have to get a lot of information from other people. Communicators in contact with response personnel at the site will be obtaining information on the extent of the damage to the truck and its contents. Transportation and radioactive material experts will be determining the likely form of the material and how much might escape. Let us assume that they tell you that they estimate that 10% of the containers may have ruptured and that most of their contents are likely to leak out within an hour or two and would quickly volatilize and become airborne. From this information we can estimate a release quantity and release time. This means we can select **Isotopic Release Rates** as the source term method.

Sample rate units: **Ci/min**

     Rationale: These units are used for convenience to make it easy to characterize the release.

Period start: **02:00**

Rationale: We were told the accident happened at 2 a.m.

Period stop: **02:10**

Rationale: Assume it took 10 minutes for all the gas to escape.

Nuclide: **H-3** and the release rate is **1500**.

Rationale: That gets all 15,000 Ci (10% of the total inventory) out in the 10 minutes.

Release Path   **Direct to atmosphere**

Rationale: The accident occurred outside with no holdup or reduction by other structures. The isotopic source term type being used assumes that the measured release rates are the rate the activity enters the atmosphere.

Release height: **0.0 m**

Consider building wake effects: **Yes**

Start of release to atmosphere: **02:00**

End of release to atmosphere: **Release duration, 0 days 00:10 hh:mm**

Meteorology   **Predefined Data (Non Site-Specific) – Winter - Night - Calm**

Rationale: Not much is known. However, it is the middle of the night and the roads are icy so winter, night calm seems likely.

Calculation Options   Distance of calculation: **Close-in only**.

End calculations: **Start of release to atmosphere + 1 hour**

Results

The source term summary confirms that all 15,000 Ci of tritium were released.

The following doses are projected:

Dose type	Distance from release (miles)			
	0.1	0.2	0.5	1.0
TEDE (rem)	3.9E-02	1.9E-02	7.6E-03	4.6E-03
Thyroid CDE (rem)	3.9E-02	1.9E-02	7.6E-03	4.6E-03

The TEDE and thyroid CDE values are well below the EPA PAGs even at 1/10 of a mile from the release point.

It seems curious that the TEDE and Thyroid CDE values are identical. Use the Nuclide Data Viewer to check the dose factors.

You will see that the dose factors for Inhalation CEDE and Thyroid are identical for H-3 (tritium). Tritium is unique in that it distributes uniformly through the body and all tissues receive the same dose. Also, tritium has no gamma, i.e. no cloud or groundshine dose, and thus the CEDE dose factor is the same as the thyroid dose factor.

The TEDE is the sum of the external dose and the inhalation CEDE. In this case, the external dose is zero. The inhalation CDE is the sum of the CDE for each organ times a weighting factor for each organ. The sum of the weighting factors is 1. And, since the tritium is evenly distributed all organs have the same CDE. Thus, the TEDE and thyroid CDE are the same.

Meteorology

This section presents a series of problems to provide more practice with the meteorological data processor and to illustrate how various meteorological conditions can affect resultant doses.

To simplify things, the following source term and release pathway will be used for all the problems:

Event Type	Nuclear Power Plant
Event Location	James A. Fitzpatrick
Source Term	Time Core is Uncovered Reactor shutdown: 03:30 Core uncovered: 06:00 Core recovered: No

Release Path	Through the dry well wall
	Not an isolated stack
	Release height: 0.0 meters
	Consider building wake effects: Yes
	Release to containment: 06:00
	Release events: all at 06:00
	  Sprays off
	  Filters off
	  Leak rate - 5% / d

35 Precipitation

Purpose

To understand how to deal with precipitation.

Discussion

Precipitation has a significant impact on projected dose. Precipitation washes particulates from the plume causing higher groundshine doses where the plume first encounters the precipitation. As the plume travels, depletion of the plume caused by precipitation will lower plume concentrations downwind. Thus, at greater distances, precipitation will cause lower cloudshine, groundshine, and inhalation doses.

RASCAL understands only seven precipitation types: none and 3 types each of liquid and frozen precipitation. The two classes are called rain and snow for simplicity. The precipitation in each class may be light, moderate, or heavy. The National Weather Service reports precipitation rates, but the format is not always uniform across the country. Rain includes drizzle, freezing rain, and freezing drizzle. Snow includes snow grains, snow pellets, ice pellets, ice crystals, and hail.

Precipitation Intensity

Precipitation intensity may be used if available. Some observations will report the rainfall rate. Forecasts normally include the predicted amount over 6 hour periods. Divide by 6 to get the hourly rate. Table 17 shows the appropriate conversion from rainfall rate to RASCAL precipitation class.

Table 17 Converting rainfall intensity to RASCAL precipitation class

Rainfall rate (intensity)	Equivalent RASCAL precipitation class
< 0.04 in/h (< 1 mm/h)	Light rain
0.04 in/h - 0.2 in/h (1 - 5 mm/h)	Moderate rain
> 0.2 in/h (> 5 mm/h)	Heavy rain

Doppler Radar

Doppler radar (NEXRAD) products display echo intensities in units of dBZ. These values correlate with precipitation type and intensity. However, the conversion varies by the area of the country and the type of precipitation that is occurring. Consult with a meteorologist if you wish to try and use these types of products as input to RASCAL.

Probability of precipitation

The probability of precipitation (PoP or POP) is described as the probability that a measurable amount of liquid precipitation (or the water equivalent of frozen precipitation) will occur. Measurable precipitation is defined as greater than 0.2 mm or 0.01 inch. The PoPs do not necessarily indicate precipitation intensity.

Forecasters often use the following worded qualifiers to express uncertainty or to qualify the area.

PoP percent	Expression of uncertainty	Equivalent areal qualifiers
10 percent	---	isolated or few
20 percent	slight chance	widely scattered
30 - 50 percent	chance	scattered
60 - 70 percent	likely	numerous
80 - 100 percent	definite	(none used)

The following is general guidance on using the PoP information in making RASCAL runs.

Probability of Precipitation (POP)	What to do in RASCAL
Slight chance of less	Set the precipitation class to "No Precip" and don't worry about it.
Chance or likely	Create two meteorological data sets, one without precipitation and the other with precipitation. Run the RASCAL scenario with both data sets and compare the doses. It is not appropriate to average the doses from the two runs. Either precipitation will occur or it won't but you will have considered both cases.
Definite	Enter the predicted precipitation

Problem

If needed, enter the initial conditions described on page 150 for source term and release path.

A. Create a meteorological data set with the following conditions:

Time	Wind direction (deg)	Wind speed (mph)	Stability class	Precipitation	Air Temp (deg F)
0600	210	6	B	none	50

Process the data and save with the new name: **FITZ No Rain**

Then, run the calculations using the following settings:

- Distance of calculation: Close-in + out to 10 miles
- End calculations at: Start of release to atmosphere plus: 6 hours

Record the **cloudshine** and **4-day groundshine** doses at 1, 5, and 10 miles in the tables provided.

Precipitation	Cloudshine (rem) at distance of:		
	1 mi	5 mi	10 mi
No rain			
Rain			

Precipitation	4-day groundshine (rem) at distance of:		
	1 mi	5 mi	10 mi
No rain			
Rain			

B. Now, lets assume that the 06:00 observation included rain. Edit the meteorological data set to change the precipitation type to **Rain**. Reprocess the data and save the data set under a new name: **FITZ Rain**.

Recalculate doses and again record the cloudshine and 4-day groundshine doses in the tables.

Results

A comparison of the 2 shine doses at the selected distances:

Precipitation	Cloudshine dose (rem) at distance of:		
	1 mi	5 mi	10 mi
No rain	8.8E-01	8.4E-02	2.8E-02
Rain	6.9E-01	4.0E-02	1.0E-02

The cloudshine doses are higher at all distances for the "no rain" case. Why?

Precipitation	4-day groundshine dose (rem) at distance of:		
	1 mi	5 mi	10 mi
No rain	5.6E+00	2.6E-01	8.2E-02
Rain	1.1E+02	1.5E+00	2.7E-02

The 4-day groundshine dose at 1 mile is higher for the rain case, but at 10 miles the 4-day groundshine for the rain case is lower. Why?

Rain washes the material out of the air and deposits it on the ground. When there is more material on the ground, groundshine dose increases. When there is less material in the air, cloudshine decreases.

If the rain does not start until after the material has moved away from the release point, "hot spots" could be created.

Summary

1. Moderate rain removes most of the non-noble gases in about an hour.

2. Rainfall decreases cloudshine doses by washing material out of the plume.

3. Rainfall increases groundshine dose close-in but decreases groundshine dose farther out.

4. RASCAL dose not model isolated precipitation events (e.g. thunderstorms) well.

36 Calm and Variable Winds

To learn how to deal with calm or variable winds.

Discussion

Calm is defined in the Glossary of Meteorology as "the absence of apparent motion of the air". It may be quantified as a wind speed of less than 1 mile per hour. It does not however mean that there is no motion of the air. Instead the air moves at low speed with lots of meander and no defined direction. Calm conditions present a problem for dispersion modeling. Since there is no defined direction and the wind speed is very low, the dispersion calculations cannot be based on plume movement. Instead, the models switch to dispersion based on time.

The 3 models within RASCAL each treat calm conditions differently. The straight-line Gaussian plume model (used close-in) and the puff model both switch to time based diffusion under calm conditions. The plume model only "sees" the winds at the release point (site) and only the winds at the time of release. The puff model (used for calculations to 10-50 miles) can encounter calm conditions anywhere within the wind field. The UF_6 plume model does *not* switch to a time-based diffusion. Instead, winds below 1 mph are treated as 1 mph. This is important to note since it means that with the UF_6 model the wind direction for a zero wind speed will be used.

Care should be taken when making protective action decisions based upon model runs using calm wind conditions. The low wind speeds and directional uncertainty cause high doses in areas not predicted by the models. In some cases, a meteorologist may be able to provide a better value (speed and direction) by examining data from surrounding weather stations.

A good approach is to go to the weather.gov tabular forecast tables. These will generally include forecast wind speed and direction data for the current time. However, this data may be up to 12 hours old. It should not be used without first contacting the local forecast office..

Problem

Use the standard release scenario (page 150) for starting this problem.

Create a new meteorological data set with the following conditions:

Time	Wind direction (deg)	Wind speed (mph)	Stability class	Precipitation	Air Temp (deg F)
0600	130	0	B	none	50

Process the data and save with the new name: **FITZCalm**

Make a calculation run using this new data set and the following settings:

- Distance of calculation: Close-in + out to 10 miles
- End calculations at: Start of release to atmosphere plus: 6 hours

Examine the thyroid CDE footprints and also record the thyroid CDE values at 0.2, 1.0, and 2.0 miles.

Met data set	Thyroid CDE (rem) at distance of:		
	0.2 mi	1.0 mi	2.0 mi
Calm			

Results

The resulting thyroid CDE values for the three cases are:

Met data set	Thyroid CDE (rem) at distance of:		
	0.2 mi	1.0 mi	2.0 mi
Calm	9.8E+02	3.9E+01	9.7E+00

Here are the thyroid CDE footprints resulting when the wind speeds were really zero. A different pattern emerges.

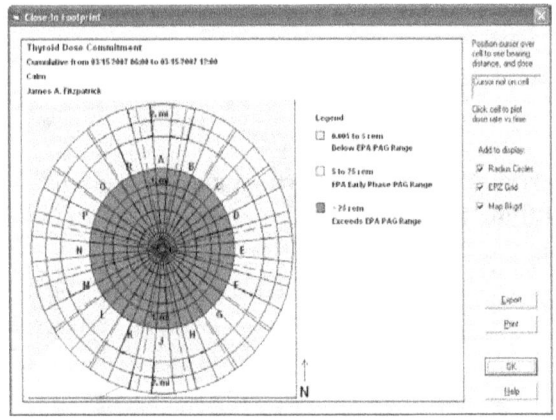 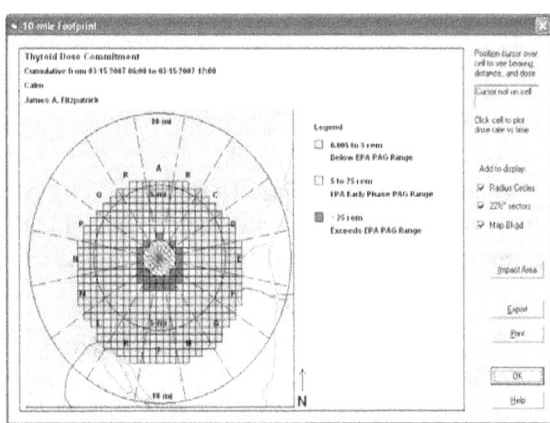

When the wind speed is less than 1 mph, diffusion coefficients are computed as a function of time. The horizontal diffusion coefficient grows at a rate of 2300 feet (700 m) per hour. The vertical diffusion coefficient grows as if the wind speed was 0.9 mph (0.4 m/s) in G stability. The material moves out in all directions from the source. Note the similarity between the plume and puff patterns.

This problem indicates the difficulty in predicting the direction of material movement under low wind speed conditions. The calm case is the extreme. Given no vector along which to move the material, the model diffuses it equally in all directions. Even with a measured direction at low speeds that direction tends to vary considerably. At low wind speeds the material can move in unexpected directions and since it is moving slowly the concentrations can be high.

37 Elevated Releases

Purpose

To see how elevated releases affect downwind inhalation doses.

Discussion

Release Height

RASCAL characterizes the release point to the atmosphere as from either a non-isolated stack or an isolated stack. For non-isolated stacks, release heights can range from ground level to 75 meters (250 feet). This option allows consideration of building wake effects. For isolated stacks, the release height range is ground level to 200 meters (650 feet). The option does not consider building wake but can consider plume rise and stack downwash.

Release Height Winds

The Meteorological Data Processor program generates wind fields for a height 10 meters above the ground. If the actual release height of the material is greater than 10 meters a wind speed profile is used to adjust the transport wind speed to the release height. This profile is dependent on atmospheric stability and surface roughness. The equation and further discussion is provided in the RASCAL technical documentation.

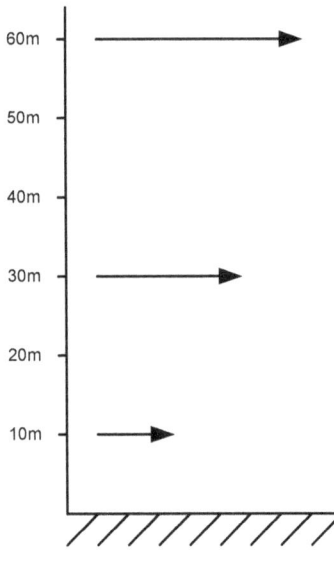

If the winds are measured at a height above 10 meters, they will be scaled down to the 10 meter level. To properly use data from elevated measurement heights you must make sure the "wind measurement height" setting for the station is set correctly.

Mixing Layer Thickness

Heating of the surface and surface friction combine to generate turbulence that mixes material released at or near ground level through a layer that varies in thickness from a few meters to a few kilometers in thickness. This layer is referred to as the mixing layer. The atmospheric models in RASCAL use the mixing height (also referred to as the mixing-layer depth or mixing-layer thickness) to limit vertical diffusion. Elevated releases within the mixing layer disperse vertically and horizontally as the material is transported by the wind. As shown in the Figure 4, the material is reflected by the top of the mixing layer and the ground surface. At some distance downwind the material becomes uniformly mixed through the layer. With elevated releases, it may be some distance downwind before the actual plume of material intersects the ground surface.

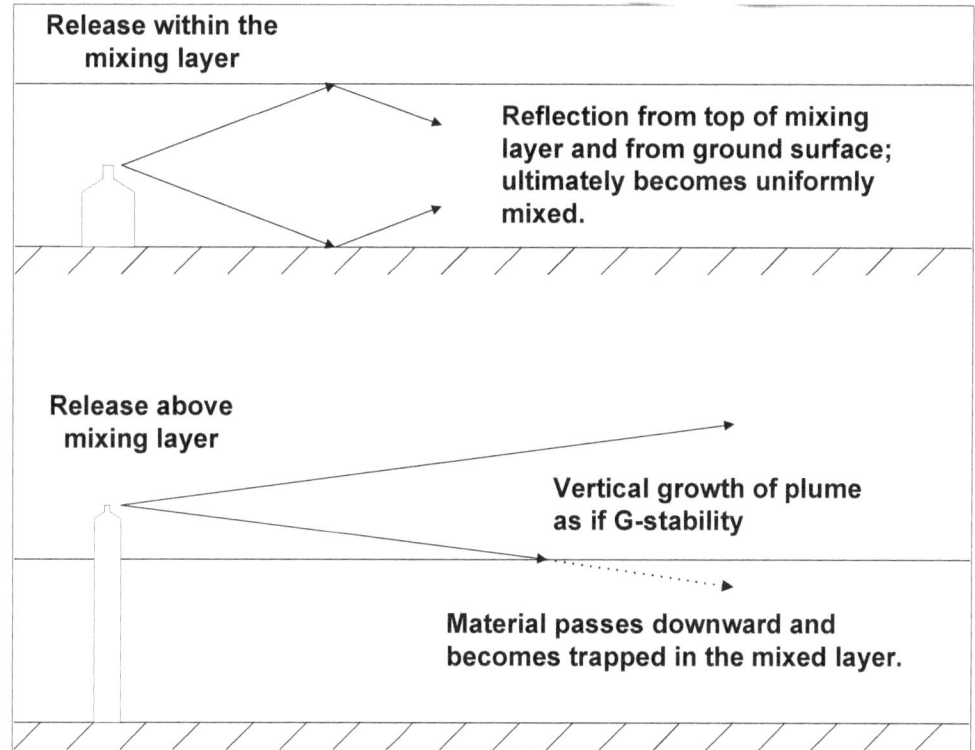

Figure 4 Plume interaction with the mixing layer

Since mixing heights are rarely measured at nuclear power plants, RASCAL provides two ways of estimating the values: from current meteorological data or from climatological data. The first is the preferred option.

Plume Rise and Stack Downwash

If an isolated stack is specified, the effective release height is the sum of three components: stack height, plume rise, and stack downwash. Plume rise is caused by buoyancy due to density differences between the effluent and the surrounding air and vertical momentum due to effluent flow rates. Both processes are modeled by RASCAL. Stack downwash occurs when the stack exit velocity is less than or equal to the wind speed at the stack exit. The downwash is an aerodynamic effect that reduces the release height by up to 3 stack diameters.

Problem

Using the standard initial conditions (Page 150) for starting these meteorology related problems, edit the release pathway to change the following:

- Release point characterization: **Isolated stack**
- Stack height: **60 meters**
- Consider plume rise: **No**

Create a new meteorological data set with the following conditions:

Time	Wind direction (deg)	Wind speed (mph)	Stability class	Precipitation	Air Temp (deg F)
0600	340	3	F	none	50

Process the data and save with the new name: **FITZ Elevated**

Make a calculation using the following settings:

.

- Distance of calculation: **Close-in + out to 25 miles**
- End calculations at: **Start of release to atmosphere plus: 6 hours**

Record the thyroid CDE and cloudshine doses in the following 2 tables in the rows labeled **No plume rise**.

Calculation	Thyroid CDE (rem) at distance of:			
	0.5 mi	2 mi	10 mi	25 mi
No plume rise				
With plume rise				

Calculation	Cloudshine (rem) at distance of:			
	0.5 mi	2 mi	10 mi	25 mi
No plume rise				
With plume rise				

Next, go back into the release pathway section and change the option to **Consider plume rise** to **Yes**. Then, click the **Edit Plume Rise Settings** button to access the **Plume Rise Parameters** screen. Enter the following data:

- Effluent flow rate = 50,000 cfm
- Stack diameter = 3 ft
- Effluent temperature = 60 °F

Repeat the calculations and again record results in the tables in the **With plume rise** rows.

The results for the two cases are shown below.

Calculation	Thyroid CDE (rem) at distance of:			
	0.5 mi	2 mi	10 mi	25 mi
No plume rise	5.8E-02	3.1E+02	6.5E+01	6.4E+00
With plume rise	< 1.0E-04	5.5E+01	5.3E+01	8.6E+00

Calculation	Cloudshine (rem) at distance of:			
	0.5 mi	2 mi	10 mi	25 mi
No plume rise	2.0E+00	1.3E+00	1.2E-01	9.8E-03
With plume rise	1.4E+00	1.0E+00	1.3E-01	1.5E-02

Questions

For the calculation with no plume rise:

1. Why are the thyroid doses at 2 miles greater than those at 0.5 mile?

2. Why are there doses at 25 miles? With a wind speed of 3 mph and 6 hours of calculation time the plume should only reach 18 miles.

For the calculation with plume rise:

3. Why are the thyroid doses higher at 25 miles than the case with no plume rise?

4. Why are the cloudshine doses lower at 0.5 and 2 miles than the case with no plume rise?

Answers to questions

1. Note that the thyroid inhalation dose at 0.5 miles is 0.058 rem. By 2 miles however this dose has risen to 310 rem. The material was released 60 meters above the ground and in the stable F stability conditions remained aloft for some distance downwind. It is only when the plume has had time to grow vertically enough to intersect the ground that it begins to contribute to inhalation dose. Notice that the cloudshine doses do not show the increase downwind.

2. Recall that the model will adjust the wind speed to the release height. In this case the wind speed of 3 mph at 10 meters above the ground was adjusted to about 5.5 mph at 60 meters above the ground. With that wind speed the material would be transported 33 miles in the 6 hour calculation period.

3. The plume rise increases the effective release height. The material stays aloft longer and is able to contribute more to the inhalation doses further downwind.

4. Again, the plume rise increased the effective release height. The material in the plume was higher above the ground at any those distances downwind and thus contributed less to the cloudshine dose.

Click the **25 mi radius gridded** button to view the wind field.

To learn how to add a meteorological station that is not in the RASCAL database.

The winds at Fitzpatrick and Syracuse have had the effect of transporting the material to the east and northeast. Our assessment might be even better if we had some actual wind data downwind of the release. After consulting a map, we see that there is the town of Watertown about 40 miles to the NE of the Fitzpatrick site. This location was not listed as one of the stations for meteorological data in the RASCAL database. However, data may be available.

If an internet connection is available, go see if you can find some current observations for Watertown, New York.

One possible way to get the data for Watertown is as follows:

1. Go to the internet site: **weather.noaa.gov**
2. In the section United States Weather, select **New York** in the drop-down box and then click the **Go** button.
3. In the section Current Weather Conditions, select **Watertown** in the list and then click the **Go** button.

An example of this type of report is shown below:

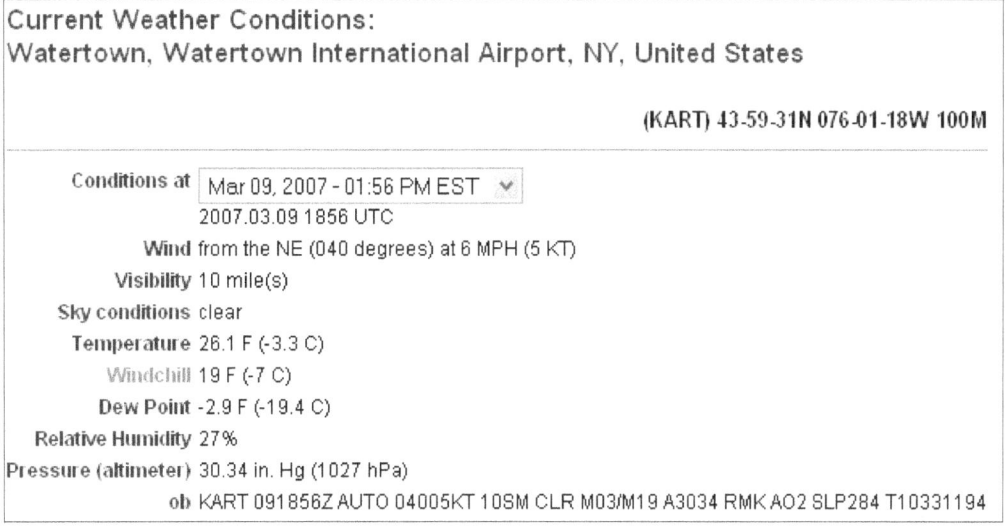

Having retrieved some current observations, we need a way to enter the data into the Meteorological Data Processor. Watertown was not listed in the sites for Fitzpatrick so it will have to be added.

Select **Stations | Add Station** from the menu bar at the top of the main screen.

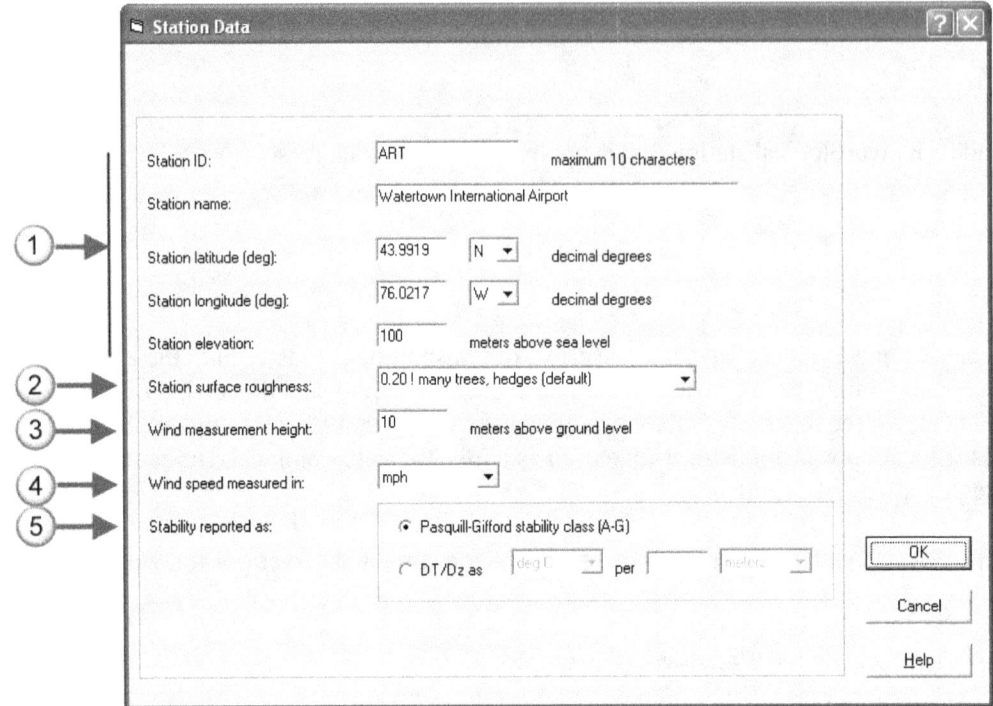

Fill in the information about the Watertown station.

1. The first two fields are used for labeling output products. They are not critical but should be accurate to avoid confusion.

    Station ID   use either the 3 or 4 letter ID code ART or KART

    Station Name   Watertown International Airport

    The station latitude and longitude are on the NWS page just below the site name. They are needed to properly locate the weather station. They are shown in degrees-minutes-seconds and need to be converted to decimal degrees (at least to the nearest 1/100 degree) before entering. Thus, the latitude in decimal degrees will be:

$$43 + \frac{59}{60} + \frac{31}{3600} = 43 + 0.9833 + 0.0086 = 43.9919$$

    Station latitude (deg)   43.9919 N

    Station longitude (deg)   76.0217 W

2. The station elevation is also shown in the NWS page after the latitude and longitude.

    Station elevation   100 meters above sea level

3. For surface roughness, unless you have some other information about the station surrounding, you should select a value from 0.1 to 0.2. This number is not critical to the calculations.

   Station surface roughness   0.20 - Many trees, hedges (default)

4. Wind measurement height should be as accurate as possible. In this case we do not know the height so enter the standard anemometer height of 10 meters.

   Wind measurement height   10 meters above ground level

5. Wind speed units can be obtained by looking at the current weather conditions report. The decoded version reports the wind speed in mph and in knots. Select the unit you will use for all the input of speed for this station.

   Wind speed measured in   mph

6. Stability is not generally reported by the NWS with station observations. In this case, just leave the selection at the default of Pasquill-Gifford stability class (A-G).

Click **OK** to return to the main screen.

If we always want Watertown to appear as a meteorological station for the Fitzpatrick site we will need to do one more step. Select **Stations | Update Station File** from the menu bar. This will write a new copy of the stations file into the Fitzpatrick data area. This new file will be used automatically in the future whenever the Fitzpatrick site is selected.

39 Stability Classes

Purpose

To learn about stability classes and how they are estimated by RASCAL.

Discussion

Stability classes are determined on the basis of quantities that can be observed. For example, there is a direct relationship between the temperature profile in the atmosphere and turbulence. As a result, NRC has used temperature structure ($\Delta T/\Delta z$) to define seven stability classes (see Table 18). Meteorological towers at nuclear power plants measure ΔT directly. However, they may report the stability class instead of the $\Delta T/\Delta z$. $\Delta T/\Delta z$ is generally reported as °C/100 m or °F/100 ft.

At some nuclear power plants, ΔT is used to estimate the stability class for the vertical dispersion parameter, and the standard deviation of wind directions, which is a direct measure of turbulence, is used to determine a second stability class specifically for the horizontal dispersion parameter. RASCAL does not have a provision for two stability classes. If two stability classes are available, the stability class for the vertical dispersion parameter should be used.

Table 18 Selection of Pasquill-Gifford stability class by $\Delta T/\Delta z$ method

Stability class	Temperature change with height **	
	(°C/100 m)	(°F/100 ft)
A	< -1.9	< -1.0
B	-1.9 to -1.7	-1.0 to -0.9
C	-1.7 to -1.5	-0.9 to -0.8
D	-1.5 to -0.5	-0.8 to -0.3
E	-0.5 to 1.5	-0.3 to 0.8
F	1.5 to 4.0	0.8 to 2.2
G	> 4.0	> 2.2

** Negative values indicate a decrease in temperature with height

The National Weather Service does not include stability class in its reports, and they generally do not have temperature profiles. Therefore methods have been developed to estimate stability classes on the basis of more readily available weather variables such as wind speed, solar radiation, and sky cover along with time of the day (see Table 19).

Table 19 Selection of Pasquill-Gifford stability class by meteorological conditions

Surface wind speed		Daytime insolation [++]			Nighttime conditions	
m/s	mph	Strong	Moderate	Slight	Thin overcast or > 50% low clouds	<= 3/8 cloudiness
< 2	< 4	A	A - B	B		
2 - 3	4 - 7	A - B	B	C	E	F
3 - 4	7 - 9	B	B - C	C	D	E
4 - 6	9 - 13	C	C - D	D	D	D
> 6	> 13	C	D	D	D	D

++ Insolation = incoming solar radiation; i.e. strength of sunlight at ground level. Use slight insolation for mostly cloudy to overcast days or when it is near dawn or dusk. Use strong insolation for no clouds at mid-day during the summer. At all other times and conditions during daytime, assume moderate insolation.

The stability can be entered into RASCAL along with the other meteorological data. It may be entered directly as a class (A to G) or as a temperature change with height ($\Delta T/\Delta z$). There is an **Unknown** selection if the stability is not known.

When the stability is unknown, RASCAL will estimate it based on the time of day, wind speed, and the precipitation entered. Table 20 shows that estimated stability class for various conditions.

Table 20 How RASCAL estimates atmospheric stability class for missing stability classes

Wind speed (m/s)	No or light precipitation	Moderate or heavy precipitation
Daytime		
≤ 6.0	C	C
> 6.0	D	D
Nighttime		
≤ 3.0	F	E
3.1 - 5.0	E	E
> 5.0	D	D

Reference: NUREG-1887, Section 6.3.2

Daytime is defined as one hour after sunrise to one hour before sunset. Nighttime is defined as one hour before sunset to one hour after sunrise. The model computes the sunrise and sunset times based upon the date and the release point latitude.

RASCAL also has the ability to adjust an entered stability class for consistency with other entered information. Table 21 shows that allowed stability class ranges for a variety of conditions. If a stability class is outside the range defined by the time of day, wind speed, and precipitation, it is set to the nearest class within the range. For example, a value of F stability during the day with winds of 6 m/s and no precipitation would be adjusted to a D class. Similarly, an A stability under the same conditions would be set to C class.

Table 21 Limits of atmospheric stability classes based on time of day, wind speed, and precipitation

Wind speed (m/s)	No or light precipitation	Moderate or heavy precipitation
Daytime		
≤ 3.0	A - E	C - E
3.1 - 5.0	B - D	C - D
> 5.0	C - D	C - D
Nighttime		
≤ 3.0	C - G	C - E
3.1 - 5.0	D - F	D - E
5.1 - 6.0	D - E	D - E
> 6.0	D	D

Reference: NUREG-1887, Section 6.3.2

40 Predefined Meteorological Data

Purpose

To learn about the predefined meteorological data available with RASCAL.

Discussion

Predefined Data (Non site-specific) is the default meteorological data set type when entering a new case. RASCAL is distributed with 18 such data sets. They were developed to cover a range of common weather conditions. These data sets have been given descriptive names such as *Summer- Afternoon - Windy*. Table 22 contains a summary of these data sets and their varying conditions.

There are pros and cons to using these predefined data sets.

Pros

- Each is a complete data set that can be used with any location and any date and time of release.
- They are quick and easy to use especially when no actual observations are available. They can be used without needing to use the Meteorological Data Processor program.

Cons

- The wind fields are uniform in space as they are based on a single data record.
- They do not consider topography
- They have a fixed wind direction of 270 degrees
- They do not include any of the diurnal variations generally seen with actual conditions.

It is *always* better to use actual meteorological data.

Table 22 Summary of predefined meteorological data sets

Data set name	Wind speed* mph	Stability class	Mixing layer* m	Precip code	Air temp deg F	Air press mb	Relative humidity %
Standard Meteorology	4	D	390	None	70	995	50
Summer - Afternoon - Calm	4	A	561	None	85	995	40
Summer - Afternoon - Rainy	8	C	884	Rain	70	995	95
Summer - Afternoon - Windy	15	B	1839	None	85	995	40
Summer - Morning - Calm	4	D	390	None	65	995	60
Summer - Morning - Rainy	8	D	780	Rain	60	995	95
Summer - Morning - Windy	15	D	1463	None	60	995	60
Summer - Night - Calm	4	F	123	None	55	995	80
Summer - Night - Rainy	8	E	333	Rain	50	995	95
Summer - Night - Windy	15	D	1463	None	50	995	85
Winter - Afternoon - Calm	4	C	442	None	35	995	75
Winter - Afternoon - Windy	20	D	1951	None	35	995	75
Winter - Morning - Calm	4	E	236	None	30	995	80
Winter - Morning - Windy	20	D	1951	None	30	995	80
Winter - Night - Calm	4	G	47	None	20	995	90
Winter - Night - Windy	20	D	1951	None	20	995	90
Winter - Rain	10	D	976	Rain	35	995	95
Winter - Snow	8	D	780	Snow	25	995	95

*Notes -   Wind direction in all cases is from the west (270 degrees).

The mixing layer depths were calculated by the Meteorological Data Processor and were not entered directly as observed data. If known, the mixing layer depth can be entered directly by the user.

Decay Calculator

Purpose

To learn how to calculate radiological decay using the Decay Calculator module.

Discussion

The decay calculator can calculate the radiological decay of a single nuclide or a group of radionuclides over any time period specified by the user.

Problem

Start with the sample from the Field Measurement to Dose problem. Assume that we want to know the concentrations that will be present one year from the time the sample was taken. Our input data is in units of $\mu Ci/m^2$, but the code can only accept activity, not activity per unit area. The way you deal with this is by mentally remembering the area units. You enter μCi, get the results in μCi, but remember that the real units are uCi/m^2.

Which radionuclides are present in significant quantities after one year?

Nuclide	Activity (µCi) at time of sample	Activity (µCi) after 1 year
Ba-140	1.4	
Cs-134	0.6	
Cs-137	0.35	
I-131	1.7	
I-132	1.3	
La-140	1.3	
Sr-89	1.0	
Te-132	280	

Enter the nuclide names and activities into the table. Set the decay interval to 1 year. Set the activity units to μCi.

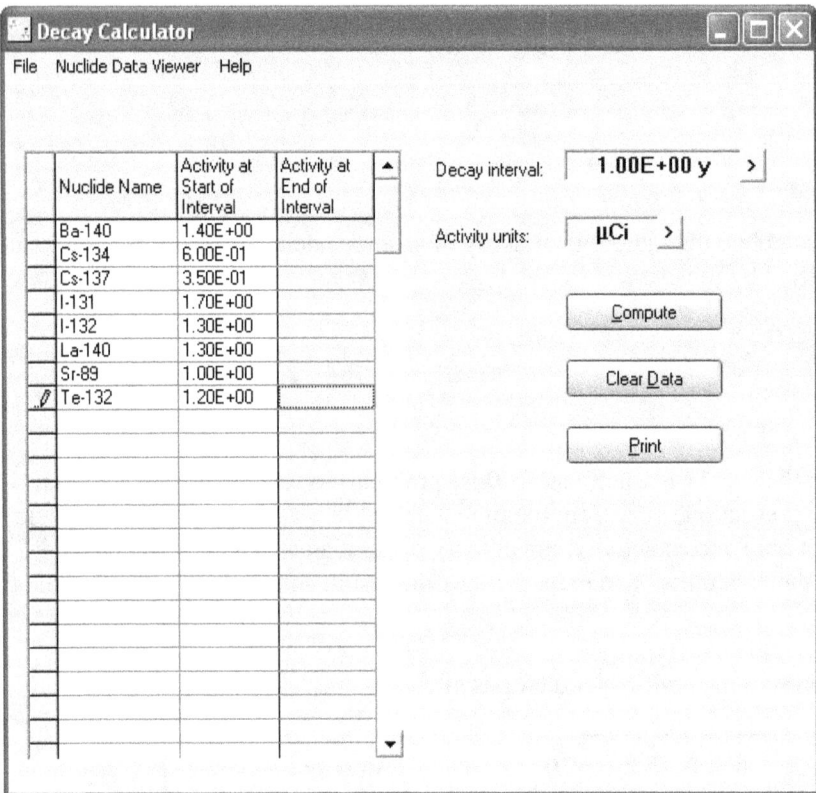

When completed, click the **Compute** button to decay the activities.

Results

The results of the 1 year of decay are displayed as shown below.

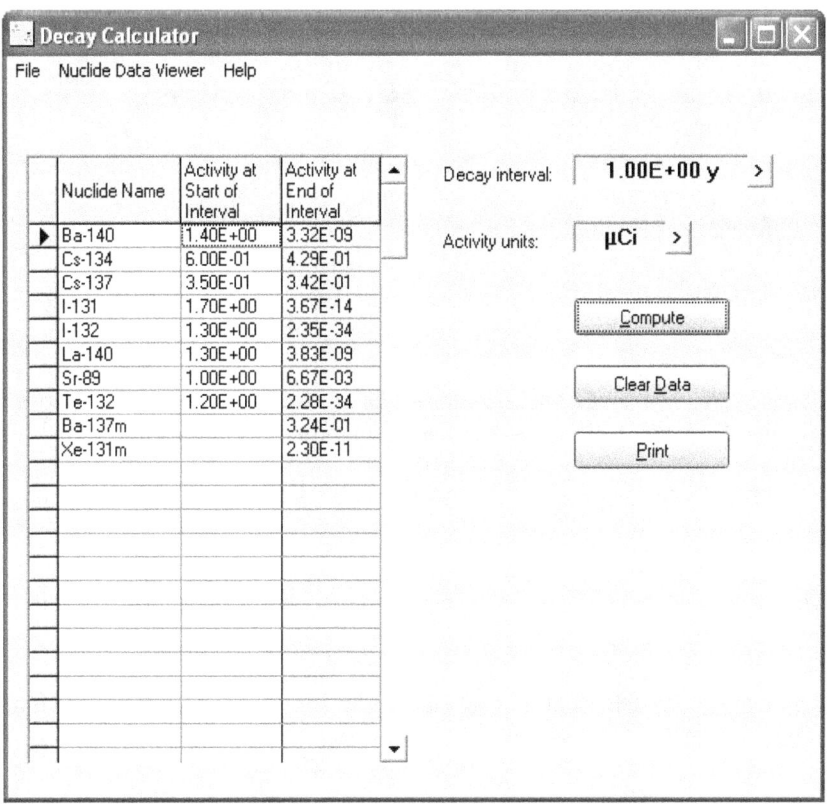

The radionuclides present in significant quantities after one year of decay are:

Nuclide	Activity (µCi)
Cs-134	0.43
Cs-137	0.34

Note that Ba-137m is also shown to be present by means of ingrowth. This is true, but not really meaningful. When a laboratory reports that a sample contains Cs-137, they mean that it contains Cs-137 + Ba-137m, the short-lived daughter of Cs-137.

References

McGuire, S.A., J. V. Ramsdell, Jr., and G. F. Athey. 2007. *RASCAL 3.0.5 Description of Models and Methods*, NUREG-1887, U.S. Nuclear Regulatory Commission.

McGuire, S.A. 1991. *Chemical Toxicity of Uranium Hexafluoride Compared to Acute Effects of Radiation*, NUREG-1391, U.S. Nuclear Regulatory Commission.

McGuire, S.A. 1988. *A Regulatory Analysis on Emergency Preparedness for Fuel Cycle and Other Radioactive Material Licensees (Final Report)*. NUREG-1140, U.S. Nuclear Regulatory Commission.

McKenna, T.J. and J. Giitter. 1988. *Source Term Estimation During Incident Response to Severe Nuclear Power Plant Accidents*, NUREG-1228, U.S. Nuclear Regulatory Commission.

Ramsdell, Jr., J.V. 1990. "Diffusion in Building Wakes for Ground-Level Releases." *Atmospheric Environment*, 24B:377-88.

Ramsdell, Jr., J.V. and C.J. Fosmire. 1998a. "Estimating Concentrations in Plumes Released in the Vicinity of Buildings: Model Development.", *Atmospheric Environment*, 32:1663-1677.

Ramsdell, Jr., J.V. and C.J. Fosmire. 1998b. "Estimating Concentrations in Plumes Released in the Vicinity of Buildings: Model Evaluation.", *Atmospheric Environment*, 32:1679-1689.

Science Applications International Corporation (SAIC). 1998. *Nuclear Fuel Cycle Facility Accident Analysis Handbook*. NUREG/CR-6410, U.S. Nuclear Regulatory Commission.

Soffer, L., et al. 1995. *Accident Source Terms for Light-Water Nuclear Power Plants, Final Report*. NUREG -1465, U.S. Nuclear Regulatory Commission.

Travis, R.J., R.E. Davis, and E.J. Grove. 1997. A Safety and Regulatory Assessment of Generic BWR and PWR Permanently Shutdown Nuclear Power Plants. NUREG/CR-6451, BNL-NUREG-52498, U.S. Nuclear Regulatory Commission.

U.S. Department of Energy (DOE). 1994. *Airborne Release Fractions/Rates and Respirable Fractions for Non-Reactor Nuclear Facilities: DOE Handbook*. Vol. 1, DOE-HDBK-3010094, U.S. Department of Energy.

U.S. Nuclear Regulatory Commission (NRC). 1975. *Reactor Safety Study: An Assessment of Accident Risks in U.S. Commercial Nuclear Power Plants*, WASH-1400 (NUREG-75/014), U.S. Nuclear Regulatory Commission.

U.S. Nuclear Regulatory Commission, Division of Incident Response Operations, Office of Nuclear Security and Incident Response. October, 2002. *Response Technical Manual*, Vol. 1, Rev. 5, U.S. Nuclear Regulatory Commission.

Appendix 1: Lessons Learned from Radiological Emergencies

This is a summary of the major lesson learned from the Chernobyl, Goiania, and other radiological emergencies. This list was adapted from work done by Tom McKenna of IAEA.

1. Failure to recognize severe emergencies

Problem: The severity of major nuclear emergencies that have occurred was not recognized or comprehended by facility operators in the initial phase even when there were indisputable indications of their severity. One reason is that severe emergencies were not considered in the preparedness process because their occurrence was considered to be inconceivable.

Solution: Emergency preparedness should address severe emergencies including those of low probability.

2. The importance of having established protective action criteria and policies

Problem: Responses to Chernobyl, Goiania, and other radiological emergencies were delayed because established criteria for taking protective actions were not in place and criteria could not be readily developed after the start of the emergency during a period of heightened emotions and mistrust of public officials and the scientific community.

Solution: Criteria and policies for taking protective actions and for the return to normalcy should be established in advance as part of the preparedness process. These criteria should be complete enough to apply to the entire conceivable range of emergency conditions and protective action decisions.

3. The destructiveness of "conservative assumptions"

Problem: The use of "conservative assumptions" led to actions that sometimes did more harm than good. Unnecessarily conservative assumptions were often used because it was not clear how to deal with uncertainties or whether existing guidance should be applied to the conditions that existed. There is a general tendency to implement actions at levels below those recommended if it is unclear whether the guidance addresses the situation at hand.

Solution: Actions should be taken based on realistic assumptions, and guidance should include clear statements on the conditions under which it applies.

4. The need for plain language explanations of decisions

Problem: Following Chernobyl, Goiania, and other emergencies, the public took inappropriate and in some cases harmful actions due to fear and misunderstandings concerning radiation risks and how to reduce them. These fears were in part due to the use of cryptic technical terms and the reluctance of technical experts to provide the definitive guidance needed and wanted by the public.

Solution: The criteria for taking protective actions should be accompanied by a plain language explanation so that the public officials who make the decisions and who inform the public of those decisions can understand the criteria and explain them to the public. The explanation must make clear what actions are appropriate and what actions are inappropriate. The explanation must explain how the recommended actions will insure the safety of the members of the public.

5. The need for planning to return to normalcy

Problem: Experience shows that after the emergency response there will be immense pressure from the public to take actions to correct the problem and return the situation to normal. Experience shows that officials, when under this intense pressure, take highly visible actions, even if those actions are only minimally effective or even counterproductive.

Solution: Guidance and plans should include a full range of post-emergency countermeasures that are both justified and optimized.

NRC FORM 335 (9-2004) NRCMD 3.7	U.S. NUCLEAR REGULATORY COMMISSION	1. REPORT NUMBER (Assigned by NRC, Add Vol., Supp., Rev., and Addendum Numbers, if any.)
	BIBLIOGRAPHIC DATA SHEET *(See instructions on the reverse)*	NUREG-1889

2. TITLE AND SUBTITLE	3. DATE REPORT PUBLISHED	
RASCAL 3.0.5 Workbook	MONTH	YEAR
	September	2007
	4. FIN OR GRANT NUMBER	
	R1110	

5. AUTHOR(S)	6. TYPE OF REPORT
George F. Athey, Athey Consulting Stephen A. McGuire, Nuclear Regulatory Commission James Van Ramsdell, Pacific Northwest National Laboratory	Technical report
	7. PERIOD COVERED *(Inclusive Dates)*
	NA

8. PERFORMING ORGANIZATION - NAME AND ADDRESS *(If NRC, provide Division, Office or Region, U.S. Nuclear Regulatory Commission, and mailing address; if contractor, provide name and mailing address.)*

Athey Consulting, P. O. Box 178, Charles Town, WV 25414-0178

Office of Nuclear Security and Incident Response, U. S. Nuclear Regulatory Commission, Washington, DC 20555-0001

Pacific Northwest National Laboratory, P. O. Box 999, Richland, WA 99352

9. SPONSORING ORGANIZATION - NAME AND ADDRESS *(If NRC, type "Same as above"; if contractor, provide NRC Division, Office or Region, U.S. Nuclear Regulatory Commission, and mailing address.)*

Office of Nuclear Security and Incident Response, U. S. Nuclear Regulatory Commission, Washington, DC 20555-0001

10. SUPPLEMENTARY NOTES

11. ABSTRACT *(200 words or less)*

The code currently used by NRC's emergency operations center for making dose projections for atmospheric releases during radiological emergencies is RASCAL version 3.0.5 (Radiological Assessment System for Consequence AnaLysis). This code was developed by NRC. RASCAL 3.0.5 evaluates releases from: nuclear power plants, spent fuel storage pools and casks, fuel cycle facilities, and radioactive material handling facilities.

The workbook contains problems designed to familiarize the user with the computer-based tools of RASCAL through hands-on problem solving. The workbook is primarily for use by students in a RASCAL training course supervised by a qualified instructor.

12. KEY WORDS/DESCRIPTORS *(List words or phrases that will assist researchers in locating the report.)*	13. AVAILABILITY STATEMENT
RASCAL, emergency response, dose assessment, dose projections, plume modeling, student workbook, training	unlimited
	14. SECURITY CLASSIFICATION
	(This Page) unclassified
	(This Report) unclassified
	15. NUMBER OF PAGES
	16. PRICE

www.ingramcontent.com/pod-product-compliance
Lightning Source LLC
Chambersburg PA
CBHW081447170526
45166CB00008B/2338